D0317904

It's another great book from CGP...

GCSE Maths isn't just about learning facts and formulas
— you'll also need to know how to use them to answer questions.

But don't panic. This book explains everything you need to know,
with plenty of worked examples and practice questions. It even includes
a **free** Online Edition you can read on your computer or tablet.

How to get your free Online Edition

Just go to **cgpbooks.co.uk/extras** and enter this code...

2964 5402 4759 6464

By the way, this code only works for one person. If somebody else has used
this book before you, they might have already claimed the Online Edition.

CGP — still the best! ☺

Our sole aim here at CGP is to produce the highest quality books —
carefully written, immaculately presented and dangerously close to being funny.

Then we work our socks off to get them out to you
— at the cheapest possible prices.

Contents

The columns for Units 1-3 above might be useful if you're studying Edexcel's 'Mathematics B' Specification (the one divided into units — ask your teacher if you're not sure which one you do). They show which units the pages cover.

Throughout this book you'll see grade stamps like these: **D F E C**
You can use these to focus your revision on easier or harder work.
But remember — to get a top grade you have to know **everything**, not just the hardest topics.

Published by CGP

Written by Richard Parsons

Updated by: Paul Jordin, Sharon Keeley-Holden, Simon Little,
Alison Palin, Andy Park, Caley Simpson, Ruth Wilbourne

With thanks to Janet Dickinson, Mark Moody and Glenn Rogers for the proofreading

ISBN: 978 1 84146 545 6

Groovy website: www.cgpbooks.co.uk
Printed by Elanders Ltd, Newcastle upon Tyne.
Jolly bits of clipart from CorelDRAW®

Photocopying — it's dull, grey and sometimes a bit naughty. Luckily, it's dead cheap,
easy and quick to order more copies of this book from CGP — just call us on 0870 750 1242. Phew!

Calculating Tips

Ah, the glorious world of GCSE Maths. OK — maybe it's more like whiffy socks at times, but learn it you must.
And there's plenty of it. Here are some nifty exam tricks that could net you quite a few lovely marks.

BODMAS   <u>B</u>rackets, <u>O</u>ther, <u>D</u>ivision, <u>M</u>ultiplication, <u>A</u>ddition, <u>S</u>ubtraction

<u>BODMAS</u> tells you the <u>ORDER</u> in which these operations should be done:
Work out <u>Brackets</u> first, then <u>Other</u> things like squaring, then <u>Divide</u> / <u>Multiply</u>
groups of numbers before <u>Adding</u> or <u>Subtracting</u> them.

EXAMPLES:

1. Work out $7 + 9 \div 3$

1) Follow BODMAS — do the <u>division</u> first... $7 + 9 \div 3$
2) ...then the <u>addition</u>: $= 7 + 3$
                                                              $= 10$

If you don't follow BODMAS, you get:
$7 + 9 \div 3$
$= 16 \div 3$
$= 5.333...$ ✗

2. Calculate $15 - 7^2$

1) The square is an 'other' so that's first: $15 - 7^2$
2) Then do the <u>subtraction</u>: $= 15 - 49$
                                                        $= -34$

3. Find $(5 + 3) \times (12 - 3)$

1) Start by working out the <u>brackets</u>: $(5 + 3) \times (12 - 3)$
2) And now the <u>multiplication</u>: $= 8 \times 9$
                                                        $= 72$

Don't Be Scared of Wordy Questions

A lot of the marks on your exam are for answering <u>wordy</u>, <u>real-life</u> questions. For these you don't
just have to do <u>the maths</u>, you've got to work out what the question's <u>asking you to do</u>.
<u>Relax</u> and work through them <u>step by step</u>.

1) <u>READ</u> the question <u>carefully</u>. Work out <u>what bit of maths</u> you need to answer it.
2) <u>Underline</u> the **INFORMATION YOU NEED** to answer the question — you might not
    have to use <u>all</u> the numbers they give you.
3) Write out the question **IN MATHS** and answer it, showing all your <u>working</u> clearly.

EXAMPLE:

A return car journey from Carlisle to Manchester uses $\frac{4}{7}$ of a tank of petrol.

It costs £56 for a <u>full tank</u> of petrol. How much does the journey cost?

1) The "$\frac{4}{7}$" tells you this is a <u>fractions</u> question. (Fractions questions are covered on page 17.)

2) You need <u>£56</u> (the cost of a full tank) and $\frac{4}{7}$ (the fraction of the tank used).
    It doesn't matter where they're driving from and to.

3) You want to know $\frac{4}{7}$ of £56, so in maths: $\frac{4}{7} \times £56 = £32$

Don't forget the units in your final answer — this is a question about cost in pounds, so the units will be £.

What's your BODMAS? About 50 kg, dude...

It's really important to check your working on BODMAS questions. You might be certain you did the
calculation right, but it's surprisingly easy to make a slip. Try this exam-style question and see how you do.

Q1   Find the value of: a) $15 - 12 \div 3$ b) $5 \times 2 + 3 \times 9$ c) $(3 + 5) \div 2 - 1$ [3 marks]

Calculating Tips

This page covers some mega-important stuff about using <u>calculators</u>.

Know Your Buttons Ⓒ

Look for these buttons on your calculator — they might be a bit different on yours.

Ans	This uses your <u>last answer</u> in your current calculation. Super useful.
$\sqrt[3]{\square}$	The <u>cube root</u> button. You might have to press <u>shift</u> first.
$S \Leftrightarrow D$	Flips your answer from a <u>fraction or root</u> to a <u>decimal</u> and vice versa.
x^{-1}	The <u>reciprocal</u> button.

The <u>RECIPROCAL</u> of a number is <u>1 DIVIDED BY IT</u>.
- So the reciprocal of 2 is ½, and the reciprocal of ¼ is 4 (1 ÷ ¼).
- <u>0</u> doesn't have a reciprocal (because you <u>can't</u> divide by 0).
- A <u>number</u> multiplied by its <u>reciprocal</u> is <u>1</u> (e.g. 2 × ½ = 1, 4 × ¼ = 1).
- <u>Dividing</u> by a <u>number</u> is the same as <u>multiplying</u> by its <u>reciprocal</u>, (i.e. ÷ 2 is the same as × ½).

BODMAS on Your Calculator Ⓓ

A BODMAS question on the <u>calculator paper</u> will be packed with <u>tricky</u> <u>decimals</u> and possibly a <u>square root</u>. You <u>could</u> do it on your calculator in one go, but that runs the risk of losing a precious mark.

EXAMPLE:

Work out $\dfrac{2.48 - 0.79}{\sqrt{9.2 + 6.35}}$.

Write down all the figures on your calculator display.

Do it in stages and write down each step:

1 Work out the number inside the <u>square root</u> sign:
[9.2] [+] [6.35] [=]

$$\frac{2.48 - 0.79}{\sqrt{9.2 + 6.35}}$$

2 Use the answer to work out the <u>bottom</u> of the fraction: [√☐] [Ans] [=]

Write the answer down and store it in the <u>memory</u> by pressing: [STO] [M+]

$$= \frac{2.48 - 0.79}{\sqrt{15.55}}$$

3 Now work out the <u>top</u> of the fraction:
[2.48] [−] [0.79] [=]

$$= \frac{1.69}{3.943348831}$$

$$= 0.4285697443$$

4 Do the division:
[1.69] [÷] [RCL] [M+] [=]

This gets the value of the <u>bottom</u> of the fraction out of the <u>memory</u>.

NOTE: On some calculators, a <u>bracket</u> opens when you use the <u>square/cube root</u> function. So to enter something like $\sqrt{12}$ + 1, you have to <u>close the bracket</u>: [√☐] [12] [)] [+] [1]

Calculators — only as clever as the button presser...

Different calculators behave differently, so get to know your own. Try everything above on your calculator.

Q1   Work out $\dfrac{\sqrt{8.67 - 4.94}}{4.21 + 8.7}$. Write down all the figures on your calculator display.   [2 marks] Ⓓ

Ordering Numbers and Place Value

Here's a nice easy page to get you going.
You need to be able to: 1) <u>Read big numbers</u>, 2) <u>Write them down</u>, 3) <u>Put numbers in order</u>.

Always Look at Big Numbers in Groups of Three (G)

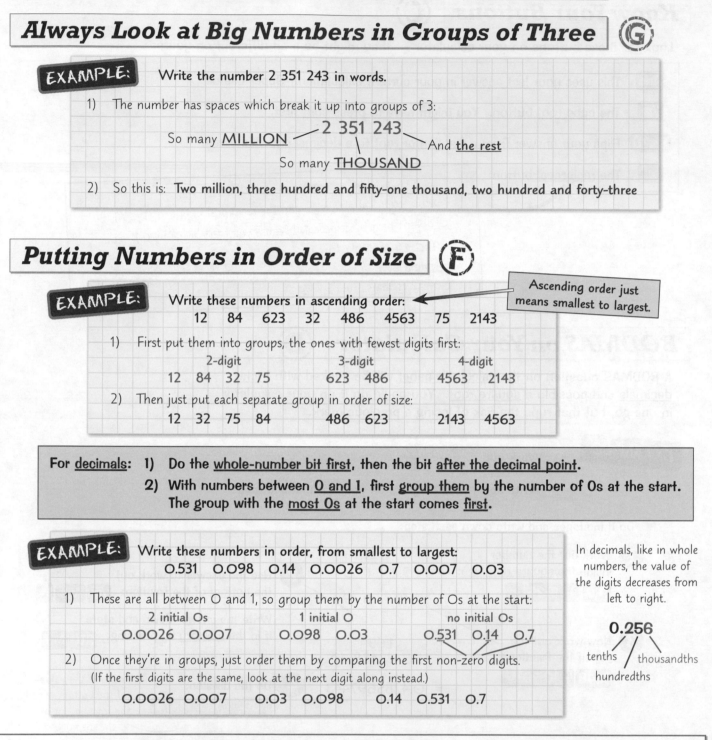

EXAMPLE: Write the number 2 351 243 in words.

1) The number has spaces which break it up into groups of 3:

So many **MILLION** — 2 351 243 — And **the rest**

So many **THOUSAND**

2) So this is: Two million, three hundred and fifty-one thousand, two hundred and forty-three

Putting Numbers in Order of Size (F)

> Ascending order just means smallest to largest.

EXAMPLE: Write these numbers in ascending order: ←

12   84   623   32   486   4563   75   2143

1) First put them into groups, the ones with fewest digits first:

2-digit	3-digit	4-digit
12 84 32 75	623 486	4563 2143

2) Then just put each separate group in order of size:

12 32 75 84	486 623	2143 4563

For <u>decimals</u>: 1) Do the <u>whole-number bit first</u>, then the bit <u>after the decimal point</u>.

2) With numbers between <u>0 and 1</u>, first <u>group them</u> by the number of 0s at the start.
The group with the <u>most 0s</u> at the start comes <u>first</u>.

EXAMPLE: Write these numbers in order, from smallest to largest:

0.531   0.098   0.14   0.0026   0.7   0.007   0.03

1) These are all between 0 and 1, so group them by the number of 0s at the start:

2 initial 0s	1 initial 0	no initial 0s
0.0026 0.007	0.098 0.03	0.531 0.14 0.7

2) Once they're in groups, just order them by comparing the first non-zero digits.
(If the first digits are the same, look at the next digit along instead.)

0.0026 0.007   0.03 0.098   0.14 0.531 0.7

> In decimals, like in whole numbers, the value of the digits decreases from left to right.
>
> **0.256**
>
> tenths / thousandths
>
> hundredths

Don't call numbers big or small to their face — they're sensitive...

There's nothing too tricky about putting numbers into order of size — just remember the tips above.
In fact, you might even find it strangely satisfying. A bit like alphabetising your CD collection.

Q1   Write these numbers in words: a) 1 234 531   b) 23 456   c) 3402   d) 203 412   [4 marks] (G)

Q2   Write this down as a number: Fifty-six thousand, four hundred and twenty-one   [1 mark] (G)

Q3   Put these numbers in order of size: 23   493   87   1029   3004   345   9   [1 mark] (G)

Q4   Write these numbers in ascending order: 0.37 0.008 0.307 0.1 0.09 0.2   [1 mark] (F)

Addition and Subtraction

With a non-calculator paper coming up, I'd imagine you'd like to learn some methods for doing sums with just a pen and paper. Well, ta-daa — here they are, just for you. Aren't I kind?

Adding (G)

1) Line up the <u>units</u> columns of each number.
2) Add up the columns from <u>right to left</u>.
3) <u>Carry over</u> any spare tens to the next column.

EXAMPLE: Add together 292, 484 and 29.

1) 292
   484
 + 29
   5
   1
   Units lined up
   2 + 4 + 9 = 15 — write 5 and carry the 1

2) 292
   484
 + 29
   05
   2 1
   9 + 8 + 2 + carried 1 = 20 — write 0 and carry the 2

3) 292
   484
 + 29
   805
   2 1
   2 + 4 + carried 2 = 8

Subtracting (G)

1) Line up the <u>units</u> columns of each number.
2) Working <u>right to left</u>, subtract the <u>bottom</u> number from the <u>top</u> number.
3) If the top number is <u>smaller</u> than the bottom number, <u>borrow</u> 10 from the left.

EXAMPLE: Work out 693 − 665.

1) 693
 − 665
   Units lined up. You can't do 3 − 5, so borrow 10 from the left.

2) 6⁸9¹³3̶
 − 665
   028
   13 − 5 = 8
   8 − 6 = 2
   6 − 6 = 0

And with Decimals... (F)

The <u>method's just the same</u>, but start instead by lining up the <u>decimal points</u>.

EXAMPLES:

1. Work out 3.74 + 24.2 + 0.6.

1) 3.74
   24.20
 + 0.60
   54
   1
   Decimal points lined up. It often helps to write in extra zeros to make all the decimals the same length
   7 + 2 + 6 = 15 — write 5 and carry the 1

2) 3.74
   24.20
 + 0.60
   28.54
   1
   3 + 4 + 0 + carried 1 = 8

2. Bob has £8, but spends 26p on chewing gum. How much is left?

1) £8.00
 − £0.26
   Decimal points lined up. 0 is smaller than 6, so you can't do 0 − 6.

2) £⁷8̶.⁰¹⁰0
 − £0.26
   Borrow 10...

3) £⁷8̶.⁹⁰¹⁰1̶00̶
 − £0.26
   £7.74
   ...then borrow 10 again
   10 − 6 = 4
   9 − 2 = 7
   7 − 0 = 7

"Carry one" — what the royals say when they want a piggyback...

Test your skills of pen-and-paper maths with these teasers:

Don't forget to include the units in your answers.

Q1 When Ric was 10 he was 142 cm tall. Since then he has grown 29 cm.
  a) How tall is he now?   b) How much more must he grow to be 190 cm tall? [2 marks] (F)

Q2 I have 3 litres of water and drink 1.28 litres. How much is left? [2 marks] (F)

Section One — Numbers

Multiplying by 10, 100, etc.

You really should know the stuff on this page because:
a) it's <u>nice 'n' simple</u>, and   b) they're likely to <u>test you on it</u> in the exam.

1) To Multiply Any Number by 10 (G)

Move the decimal point <u>ONE</u> place <u>BIGGER</u> and if it's needed, <u>ADD A ZERO</u> on the end.

E.g.   23.6 × 10 = 2 3 6

485 × 10 = 4 8 5 0

45.678 × 10 = 4 5 6 . 7 8

2) To Multiply Any Number by 100 (G)

Move the decimal point <u>TWO</u> places <u>BIGGER</u> and <u>ADD ZEROS</u> if necessary.

E.g.   296.5 × 100 = 2 9 6 5 0

34 × 100 = 3 4 0 0

2.543 × 100 = 2 5 4 . 3

3) To Multiply by 1000 or 10 000, *the same rule applies:* (F)

Move the decimal point so many places <u>BIGGER</u> and <u>ADD ZEROS</u> if necessary.

E.g.   341 × 1000 = 3 4 1 0 0 0

2.3542 × 10 000 = 2 3 5 4 2

You always <u>move</u> the <u>DECIMAL POINT</u> this much:
<u>1 place for 10</u>,   <u>2 places for 100</u>,
<u>3 places for 1000</u>,   <u>4 for 10 000</u>   etc.

4) To Multiply by Numbers like 20, 300, 8000 etc. (E)

<u>MULTIPLY</u> by <u>2</u> or <u>3</u> or <u>8</u> etc. <u>FIRST</u>, then move the decimal point so many places <u>BIGGER</u> (↪) according to how many noughts there are.

EXAMPLE: Calculate 234 × 200.

1)   First multiply by 2...   234 × 2 = 468
2)   ...then move the decimal point 2 places.   468 × 100 = 46800

Adding zeros when they're not needed? Tut, tut, noughty, noughty...

Learn the multiplying methods on this page — nothing too strenuous. For a bit of a workout, try these:

Q1   Work out   a) 12.3 × 100   b) 345 × 10   c) 9.65 × 1000   [3 marks] (F)

Q2   Work out   a) 2.4 × 20   b) 1.5 × 300   c) 60 × 3000   [3 marks] (E)

Dividing by 10, 100, etc.

This is <u>pretty easy</u> stuff too. Just <u>make sure you know it</u> — that's all.

1) To Divide Any Number by 10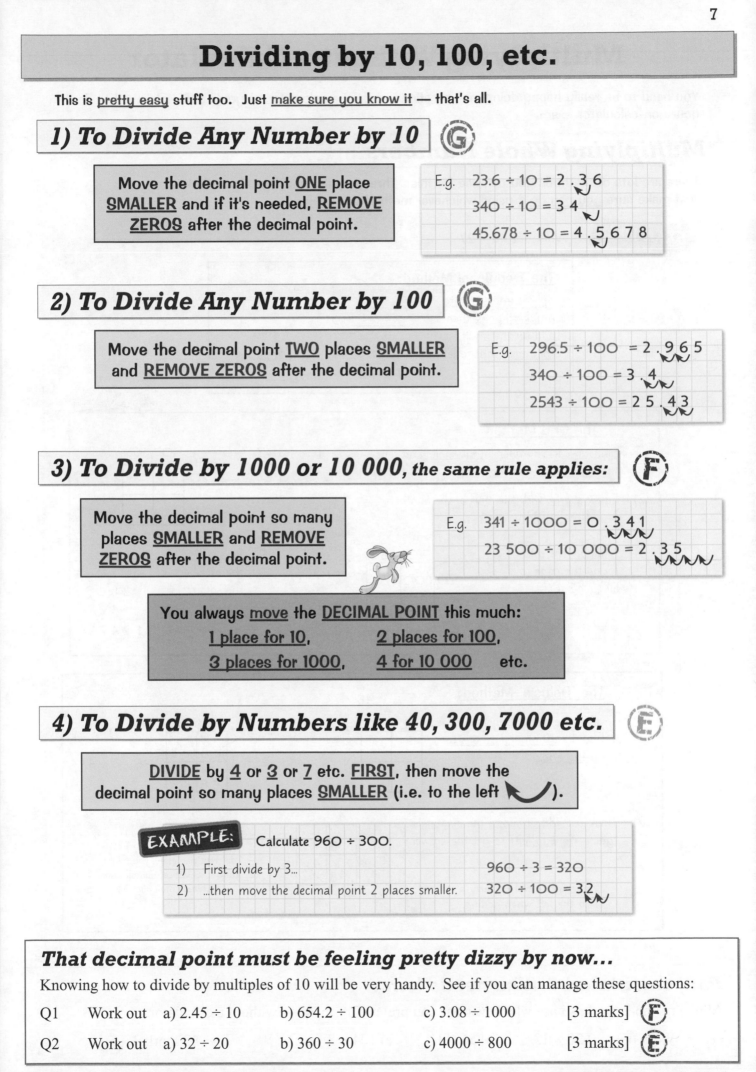

Move the decimal point <u>ONE</u> place <u>SMALLER</u> and if it's needed, <u>REMOVE ZEROS</u> after the decimal point.

E.g. $23.6 \div 10 = 2.36$
$340 \div 10 = 34$
$45.678 \div 10 = 4.5678$

2) To Divide Any Number by 100

Move the decimal point <u>TWO</u> places <u>SMALLER</u> and <u>REMOVE ZEROS</u> after the decimal point.

E.g. $296.5 \div 100 = 2.965$
$340 \div 100 = 3.4$
$2543 \div 100 = 25.43$

3) To Divide by 1000 or 10 000, *the same rule applies:*

Move the decimal point so many places <u>SMALLER</u> and <u>REMOVE ZEROS</u> after the decimal point.

E.g. $341 \div 1000 = 0.341$
$23\,500 \div 10\,000 = 2.35$

You always <u>move</u> the <u>DECIMAL POINT</u> this much:
<u>1 place for 10</u>,   <u>2 places for 100</u>,
<u>3 places for 1000</u>,   <u>4 for 10 000</u>   etc.

4) To Divide by Numbers like 40, 300, 7000 etc.

<u>DIVIDE</u> by <u>4</u> or <u>3</u> or <u>7</u> etc. <u>FIRST</u>, then move the decimal point so many places <u>SMALLER</u> (i.e. to the left).

EXAMPLE: Calculate $960 \div 300$.

1) First divide by 3... $960 \div 3 = 320$
2) ...then move the decimal point 2 places smaller. $320 \div 100 = 3.2$

That decimal point must be feeling pretty dizzy by now...

Knowing how to divide by multiples of 10 will be very handy. See if you can manage these questions:

Q1   Work out   a) $2.45 \div 10$ b) $654.2 \div 100$ c) $3.08 \div 1000$ [3 marks] **(F)**

Q2   Work out   a) $32 \div 20$ b) $360 \div 30$ c) $4000 \div 800$ [3 marks] **(E)**

Multiplying Without a Calculator

You need to be really happy doing multiplications <u>without</u> a calculator — you'll definitely need to do it in your non-calculator exam.

Multiplying Whole Numbers (E)

There are lots of methods you can use for this. Three popular ones are shown below.
Just make sure <u>you can do it</u> using whichever method <u>you prefer</u>...

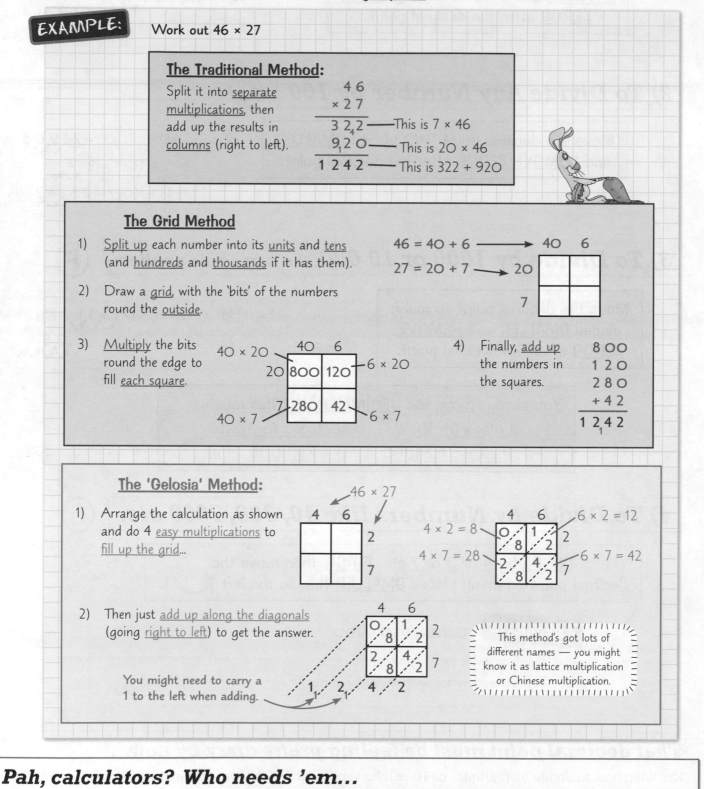

EXAMPLE: Work out 46 × 27

The Traditional Method:

Split it into <u>separate multiplications</u>, then add up the results in <u>columns</u> (right to left).

```
    4 6
  × 2 7
  3 2 2  ——— This is 7 × 46
  9 2 O  ——— This is 20 × 46
  1 2 4 2  ——— This is 322 + 920
```

The Grid Method

1) <u>Split up</u> each number into its <u>units</u> and <u>tens</u> (and <u>hundreds</u> and <u>thousands</u> if it has them).

2) Draw a <u>grid</u>, with the 'bits' of the numbers round the <u>outside</u>.

3) <u>Multiply</u> the bits round the edge to fill <u>each square</u>.

46 = 40 + 6
27 = 20 + 7

```
        40    6
40 × 20  ┌────┬────┐
    20   │800 │120 │ ← 6 × 20
         ├────┼────┤
     7   │280 │ 42 │
         └────┴────┘
40 × 7        6 × 7
```

4) Finally, <u>add up</u> the numbers in the squares.

```
  8 O O
  1 2 O
  2 8 O
+   4 2
  1 2 4 2
```

The 'Gelosia' Method:

1) Arrange the calculation as shown and do 4 <u>easy multiplications</u> to <u>fill up the grid</u>...

46 × 27

4 × 2 = 8
4 × 7 = 28
6 × 2 = 12
6 × 7 = 42

2) Then just <u>add up along the diagonals</u> (going <u>right to left</u>) to get the answer.

You might need to carry a 1 to the left when adding.

This method's got lots of different names — you might know it as lattice multiplication or Chinese multiplication.

Pah, calculators? Who needs 'em...

All the methods work, so use whichever one you prefer. Try these — without a calculator.

Q1 Work out a) 28 × 12   b) 56 × 11   c) 104 × 8   [3 marks] (E)

Dividing Without a Calculator

OK, time for <u>dividing</u> now. If you don't learn these <u>basic methods</u>, you could find yourself stuck in the exam...

Dividing Whole Numbers (E)

There are two common ways to do <u>division</u> — <u>long division</u> and <u>short division</u>.
Here's an example done using <u>both methods</u> so you can compare them. <u>Learn</u> the method you find easier.

EXAMPLE: What is 748 ÷ 22?

Short Division

number you're dividing by

number you're dividing

22 | 7 4 8

1) Set out the division as shown.

2) Look at the first digit under the line.
   7 doesn't divide by 22, so <u>put a zero</u>
   above and look at the <u>next digit</u>.

   O
   22 | 7 4 8

3) 22 × 3 = 66, so 22 into 74 goes <u>3 times</u>,
   with a <u>remainder</u> of 74 − 66 = 8.

   carry the remainder

   O 3
   22 | 7 4 88 8

4) 22 into 88 goes <u>4 times exactly</u>.

   the top line has
   the final answer

   O 3 4
   22 | 7 4 88

So 748 ÷ 22 = 34

> For questions like this,
> it's useful to write out the
> first few multiples of the
> number you're dividing by,
> e.g.   1 × 22 = 22
>   2 × 22 = 44
>   3 × 22 = 66
>   4 × 22 = 88
>   5 × 22 = 110...

Long Division

1) Set out the division as shown.

   22 | 7 4 8

2) 7 doesn't divide by 22. <u>Write a zero</u>
   above the 7 and look at the <u>next digit</u>.

   O 3 4
   22 | 7 4 8

3) 22 into 74 goes <u>3 times</u>, so put a <u>3</u> above the 4.

   − 6 6

4) <u>Take away</u> 3 × 22 = 66 from 74.
   Write the answer <u>underneath</u>, and
   move the digits after the 74 down too.

   8 8
   − 8 8

5) 22 into 88 goes <u>4 times</u>, so put a <u>4</u> above the 8.
   <u>Take away</u> 4 × 22 = 88 from 88.
   That leaves O, so we're done.

   O

   So 748 ÷ 22 = 34

Longing for some division questions? They'll be here shortly...

Think you know how to tackle division questions? OK, let's see what you're made of...

Q1   Work out   a) 96 ÷ 8   b) 84 ÷ 7   c) 252 ÷ 12   [3 marks] (E)

Q2   Joey has a plank of wood which is 220 cm long.
      He cuts it into 14 cm pieces. What length of wood will he have left over?   [2 marks] (E)

Multiplying and Dividing with Decimals

You might get a nasty non-calculator question on multiplying or dividing using decimals. Luckily, these aren't really any harder than the whole-number versions. You just need to know what to do in each case.

Multiplying Decimals (D)

1) Start by <u>ignoring</u> the decimal points. Do the multiplication using <u>whole numbers</u>.

2) Count the <u>total</u> number of digits after the <u>decimal points</u> in the original numbers.

3) Make the answer have the <u>same number</u> of decimal places.

EXAMPLE: Work out 4.6 × 2.7

We know this 'cos we worked it out on page 8.

1) Do the whole-number multiplication: 46 × 27 = 1242

2) Count the digits after the decimal points: 4.<u>6</u> × 2.<u>7</u> has <u>2 digits</u> after the decimal points, so the answer will have 2 digits after the decimal point.

3) Give the answer the same number of decimal places: 4.6 × 2.7 = **12.42**

Dividing a Decimal by a Whole Number (D)

For these, you just set the question out like a whole-number division <u>but</u> put the <u>decimal point</u> in the answer <u>right above</u> the one in the question.

EXAMPLE: What is 52.8 ÷ 3?

Put the decimal point in the answer above the one in the question

$$3 \overline{)5^2 2 . 8} \quad\quad 1 . $$

3 into 5 goes once, carry the remainder of 2

$$3 \overline{)5^2 2 .^1 8} \quad\quad 1 7 . $$

3 into 22 goes 7 times, carry the remainder of 1

$$3 \overline{)5^2 2 .^1 8} \quad\quad 1 7 . 6 $$

3 into 18 goes 6 times exactly

So 52.8 ÷ 3 = **17.6**

Dividing a Number by a Decimal (D)

Two-for-one here — this works if you're dividing a whole number by a decimal, or a decimal by a decimal.

EXAMPLE: What is 36.6 ÷ 0.12?

1) The trick here is to write it as a fraction: $36.6 \div 0.12 = \dfrac{36.6}{0.12}$

2) Get rid of the decimals by multiplying top and bottom by 100 (see p6): $= \dfrac{3660}{12}$

3) It's now a decimal-free division that you know how to solve:

$$12 \overline{)3 ^3 6 6 0} \quad\quad 0 $$

12 into 3 won't go so carry the 3

$$12 \overline{)3 ^3 6 6 0} \quad\quad 0 3 $$

12 into 36 goes 3 times exactly

$$12 \overline{)3 ^3 6 ^6 0} \quad\quad 0 3 0 $$

12 into 6 won't go so carry the 6

$$12 \overline{)3 ^3 6 ^6 0} \quad\quad 0 3 0 5 $$

12 into 60 goes 5 times exactly

So 36.6 ÷ 0.12 = **305**

The decimals came in two by two...

Just like the whole-number calculations on p8-9, use the method you prefer for multiplying or dividing.

Q1   Work out   a) 3.2 × 56   b) 0.6 × 10.2   c) 5.5 × 10.2   [3 marks] (D)

Q2   Calculate   a) 33.6 ÷ 0.6   b) 45 ÷ 1.5   c) 84.6 ÷ 0.12   [3 marks] (D)

Negative Numbers

Numbers less than zero are <u>negative</u>. You should be able to <u>add</u>, <u>subtract</u>, <u>multiply</u> and <u>divide</u> with them.

Adding and Subtracting with Negative Numbers (F)

Use the <u>number line</u> for <u>addition</u> and <u>subtraction</u> involving negative numbers:

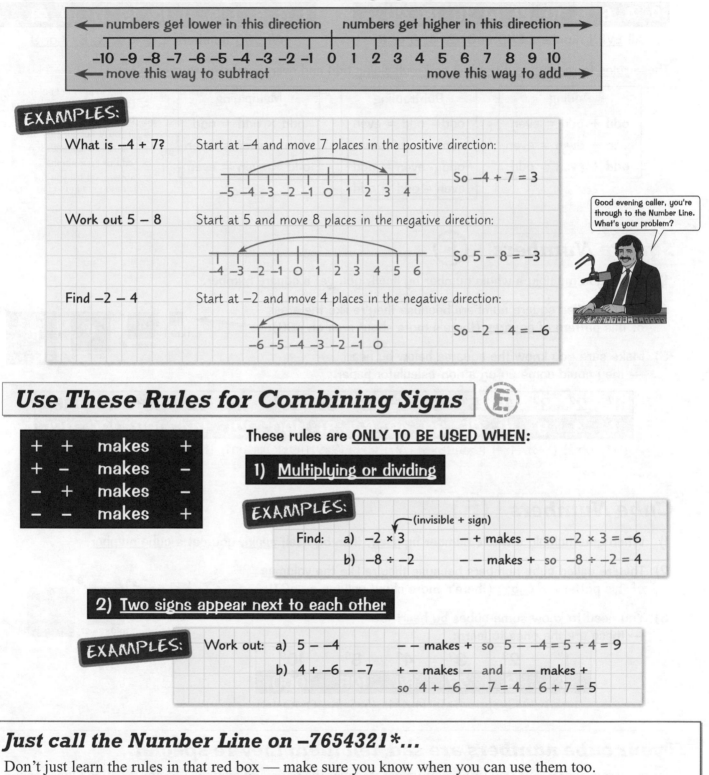

← numbers get lower in this direction   numbers get higher in this direction →

–10 –9 –8 –7 –6 –5 –4 –3 –2 –1 0 1 2 3 4 5 6 7 8 9 10

← move this way to subtract   move this way to add →

EXAMPLES:

What is –4 + 7?   Start at –4 and move 7 places in the positive direction:

So –4 + 7 = 3

Work out 5 – 8   Start at 5 and move 8 places in the negative direction:

So 5 – 8 = –3

Find –2 – 4   Start at –2 and move 4 places in the negative direction:

So –2 – 4 = –6

Good evening caller, you're through to the Number Line. What's your problem?

Use These Rules for Combining Signs (E)

+	+	makes	+
+	–	makes	–
–	+	makes	–
–	–	makes	+

These rules are **ONLY TO BE USED WHEN:**

1) Multiplying or dividing

EXAMPLES:

Find: a) –2 × 3 *(invisible + sign)*   – + makes – so –2 × 3 = –6

b) –8 ÷ –2   – – makes + so –8 ÷ –2 = 4

2) Two signs appear next to each other

EXAMPLES: Work out: a) 5 – –4   – – makes + so 5 – –4 = 5 + 4 = 9

b) 4 + –6 – –7   + – makes – and – – makes +
so 4 + –6 – –7 = 4 – 6 + 7 = 5

Just call the Number Line on –7654321...*

Don't just learn the rules in that red box — make sure you know when you can use them too.

Q1   The temperature in Mathchester at 9 am on Monday was 4 °C.
At 9 am on Tuesday the temperature was –2 °C.
a)   What was the change in temperature from Monday to Tuesday?   [1 mark] (F)
b)   The temperature at 9 am on Wednesday was 3 °C lower than on Tuesday.
What was the temperature on Wednesday?   [1 mark] (F)

*Don't really. Even if your phone accepts negative numbers.

Special Types of Number

You need to know all the types of number on this page. They're each <u>special</u> in their very own way. Bless.

Even and Odd Numbers (F)

EVEN numbers all divide by 2

| 2 | 4 | 6 | 8 | 10 | 12 | 14 | 16 | 18 | 20 ... |

All <u>EVEN</u> numbers <u>END</u> in <u>0, 2, 4, 6 or 8</u>

ODD numbers don't divide by 2

| 1 | 3 | 5 | 7 | 9 | 11 | 13 | 15 | 17 | 19 | 21 ... |

All <u>ODD</u> numbers <u>END</u> in <u>1, 3, 5, 7 or 9</u>

These <u>rules</u> for <u>adding, subtracting and multiplying</u> odd and even numbers are <u>always true</u>:

Adding	Subtracting	Multiplying
odd + odd = even	odd − odd = even	odd × odd = odd
even + even = even	even − even = even	even × even = even
odd + even = odd	odd − even = odd	odd × even = even
	even − odd = odd	

Don't stress too hard trying to remember these rules — if you're not sure, try doing a calculation with some odd or even numbers. The answer will tell you the rule.

Square Numbers (F)

1) When you <u>multiply</u> a whole number by <u>itself</u>, you get a <u>square number</u>.

2) They're called <u>square</u> numbers because they're like the <u>areas</u> of this pattern of <u>squares</u> (there's more about area on p63):

3) Make sure you know the squares below <u>by heart</u>
— they could come up on a non-calculator paper.

1^2	2^2	3^2	4^2	5^2	6^2	7^2	8^2	9^2	10^2	11^2	12^2	13^2	14^2	15^2
1	4	9	16	25	36	49	64	81	100	121	144	169	196	225

(1×1) (2×2) (3×3) (4×4) (5×5) (6×6) (7×7) (8×8) (9×9) (10×10) (11×11) (12×12) (13×13) (14×14) (15×15)

Cube Numbers (E)

1) When you <u>multiply</u> a whole number by <u>itself</u>, then by itself <u>again</u>, you get a <u>cube number</u>.

2) They're called <u>cube</u> numbers because they're like the <u>volumes</u> of this pattern of <u>cubes</u> (there's more about volume on p67):

3) You need to know some cubes <u>by heart</u> too
— these are the ones to learn:

1^3	2^3	3^3	4^3	5^3	10^3
1	8	27	64	125	1000

(1×1×1) (2×2×2) (3×3×3) (4×4×4) (5×5×5) (10×10×10)

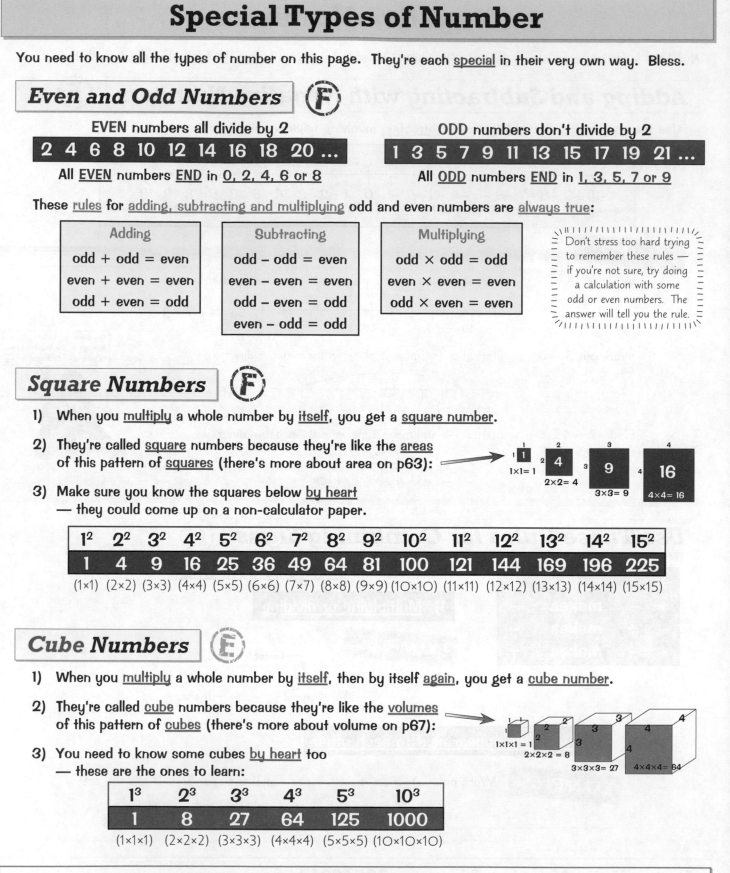

If your cube numbers are sad, tell them they're special...

A lot to take in there — if you think you've got it, cover the page and try these Exam Practice Questions:

Q1   Look at these numbers: 23, 125, 45, 32, 56, 81, 30. From this list, find:
      a) an odd number less than 40   b) a cube number   [2 marks] (E)

Q2   If a is an odd number, is $a^2 - 2$ always odd, always even or
      sometimes odd and sometimes even? Explain your answer.   [2 marks] (E)

Prime Numbers

There's one more special number sequence you need to know about — the <u>prime numbers</u>...

PRIME Numbers Don't Divide by Anything (E)

<u>Prime numbers</u> are all the numbers that <u>DON'T</u> come up in <u>times tables</u>:

| 2 | 3 | 5 | 7 | 11 | 13 | 17 | 19 | 23 | 29 | 31 | 37 | ... |

The <u>only way</u> to get <u>ANY PRIME NUMBER</u> is: 1 × ITSELF

E.g.   The <u>only</u> numbers that multiply to give 7 are   1 × 7
         The <u>only</u> numbers that multiply to give 31 are 1 × 31

> **EXAMPLE:**   Show that 24 is not a prime number.
>
> Just find another way to make 24 other than 1 × 24:   2 × 12 = 24
>
> 24 divides by other numbers apart from 1 and 24, so it isn't a prime number.

Five Important Facts

1) 1 is <u>NOT</u> a prime number.
2) 2 is the <u>ONLY</u> even prime number.
3) The first four prime numbers are <u>2, 3, 5 and 7</u>.
4) <u>Prime numbers</u> end in <u>1, 3, 7 or 9</u> (2 and 5 are the only exceptions to this rule).
5) But <u>NOT ALL</u> numbers ending in <u>1, 3, 7 or 9</u> are primes, as shown here:
   (Only the <u>circled ones</u> are <u>primes</u>.)

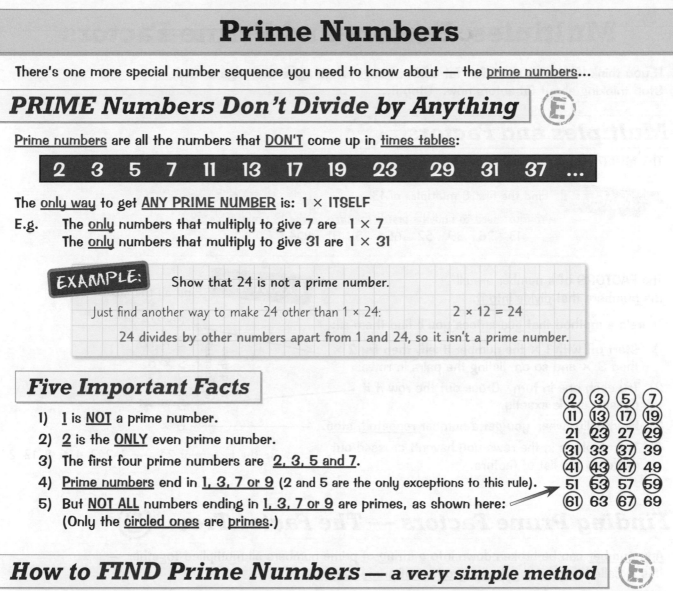

How to FIND Prime Numbers — *a very simple method* (E)

1) <u>All primes</u> (above 5) <u>end in 1, 3, 7 or 9</u>. So ignore any numbers that don't end in one of those.
2) Now, to find which of them <u>ACTUALLY ARE</u> primes you only need to <u>divide each one by 3 and 7</u>. If it doesn't divide exactly by either 3 or 7 then it's a prime.

This simple rule <u>using just 3 and 7</u> is true for checking primes <u>up to 120</u>.

> **EXAMPLE:**   Find all the prime numbers in this list: 71, 72, 73, 74, 75, 76, 77, 78
>
> **1** First, get rid of anything that doesn't end in 1, 3, 7 or 9:   71, ~~72~~, 73, ~~74~~, ~~75~~, ~~76~~, 77, ~~78~~
>
> **2** Now try dividing 71, 73 and 77 by 3 and 7:
> | | |
> |---|---|
> | 71 ÷ 3 = 23.667 | 71 ÷ 7 = 10.143   so 71 is a prime number |
> | 73 ÷ 3 = 24.333 | 73 ÷ 7 = 10.429   so 73 is a prime number |
> | 77 ÷ 3 = 25.667 BUT: | 77 ÷ 7 = 11 — 11 is a whole number, so 77 is NOT a prime, because it divides by 7. |
>
> So the prime numbers in the list are 71 and 73.

Two is the oddest prime of all — it's the only one that's even...

Learn all three sections above, then cover the page and try this Exam Practice Question without peeking:

Q1   Below is a list of numbers. Write down all the prime numbers from the list.
         39, 51, 46, 35, 61, 53, 42, 47
                                                                    [1 mark] (E)

Multiples, Factors and Prime Factors

If you think 'factor' is short for 'fat actor', I suggest you give this page a read.
Stop thinking about fat actors now. Stop it...

Multiples and Factors (E)

The MULTIPLES of a number are just its <u>times table</u>.

> **EXAMPLE:** Find the first 8 multiples of 13.
> You just need to find the first 8 numbers in the 13 times table:
> 13   26   39   52   65   78   91   104

The FACTORS of a number are all the numbers that <u>divide into it</u>.

There's a method that guarantees you'll find them all:

1) **Start off with 1 × the number itself**, then try **2 ×**, then **3 ×** and so on, listing the pairs in rows.

2) **Try each one in turn.** Cross out the row if it doesn't divide exactly.

3) **Eventually, when you get a number <u>repeated</u>, <u>stop</u>.**

4) **The numbers in the rows you haven't crossed out make up the list of factors.**

> **EXAMPLE:** Find all the factors of 24.
>
> 1 × 24
> 2 × 12
> 3 × 8
> 4 × 6
> ~~5 ×~~
> 6 × 4
>
> *Increasing by 1 each time*
>
> So the <u>factors of 24</u> are: 1, 2, 3, 4, 6, 8, 12, 24

Finding Prime Factors — The Factor Tree (C)

<u>Any number</u> can be broken down into a string of prime numbers all multiplied together — this is called '<u>expressing it as a product of prime factors</u>'.

> **EXAMPLE:** Express 420 as a product of prime factors.
>
>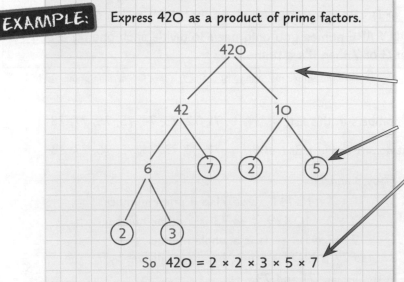
>
> So 420 = 2 × 2 × 3 × 5 × 7

To write a number as a product of its prime factors, use the mildly entertaining <u>Factor Tree</u> method:

1) **Start with the number at the top,** and <u>split</u> it into <u>factors</u> as shown.

2) **Every time you get a prime, <u>ring it</u>.**

3) **Keep going until you can't go further** (i.e. you're just left with primes), then write the primes out <u>in order</u>.

Takes me back, scrumping prime factors from the orchard...

Make sure you know the Factor Tree method inside out, then give these Exam Practice Questions a go...

Q1   Look at this list of numbers: 11, 12, 13, 14, 15, 16, 17, 18
　　　From the list, find:　　a) a multiple of 5　　b) a factor of 39　　[2 marks] (E)

Q2   Express 160 as a product of its prime factors.　　[2 marks] (C)

LCM and HCF

Two big fancy names but don't be put off — they're both <u>real easy</u>.

LCM — 'Lowest Common Multiple' Ⓒ

'<u>Lowest Common Multiple</u>' — sure, it sounds kind of complicated, but all it means is this:

> The <u>SMALLEST</u> number that will <u>DIVIDE BY ALL</u> the numbers in question.

METHOD:
1) <u>LIST</u> the <u>MULTIPLES</u> of <u>ALL</u> the numbers.
2) Find the <u>SMALLEST</u> one that's in <u>ALL the lists</u>.
3) Easy peasy innit?

The LCM is sometimes called the Least (instead of 'Lowest') Common Multiple.

EXAMPLE: Find the lowest common multiple (LCM) of 12 and 15.

Multiples of 12 are: 12, 24, 36, 48, (60,) 72, 84, 96, ...
Multiples of 15 are: 15, 30, 45, (60,) 75, 90, 105, ...

So the <u>lowest common multiple</u> (LCM) of 12 and 15 is 60.
Told you it was easy.

HCF — 'Highest Common Factor' Ⓒ

'<u>Highest Common Factor</u>' — all it means is <u>this</u>:

> The <u>BIGGEST</u> number that will <u>DIVIDE INTO ALL</u> the numbers in question.

METHOD:
1) <u>LIST</u> the <u>FACTORS</u> of <u>ALL</u> the numbers.
2) Find the <u>BIGGEST</u> one that's in <u>ALL the lists</u>.
3) Easy peasy innit?

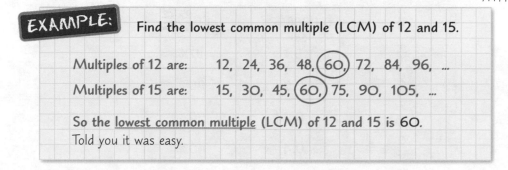

EXAMPLE: Find the highest common factor (HCF) of 36, 54, and 72.

Factors of 36 are: 1, 2, 3, 4, 6, 9, 12, (18,) 36
Factors of 54 are: 1, 2, 3, 6, 9, (18,) 27, 54
Factors of 72 are: 1, 2, 3, 4, 6, 8, 9, 12, (18,) 24, 36, 72

So the <u>highest common factor</u> (HCF) of 36, 54 and 72 is 18.
Told you it was easy.

Just <u>take care</u> listing the factors — make sure you use the <u>proper method</u> (as shown on the previous page) or you'll miss one and blow the whole thing out of the water.

LCM and HCF live together — it's a House of Commons...

You need to learn what LCM and HCF are, and how to find them. Turn over and write it all down. And after that, some lovely Exam Practice Questions — bonus.

Q1 Find the lowest common multiple (LCM) of 9 and 12. [2 marks] Ⓒ

Q2 Find the highest common factor (HCF) of 36 and 84. [2 marks] Ⓒ

Fractions, Decimals and Percentages

Fractions, decimals and percentages are <u>three different ways</u> of describing when you've got <u>part</u> of a <u>whole thing</u>. They're <u>closely related</u> and you can <u>convert between them</u>.

This table shows the really common conversions which you should know straight off without having to work them out:

Fractions with a 1 on the top (e.g. $\frac{1}{2}$, $\frac{1}{3}$, $\frac{1}{4}$, etc.) are called <u>unit fractions</u>.

Fraction	Decimal	Percentage
$\frac{1}{2}$	0.5	50%
$\frac{1}{4}$	0.25	25%
$\frac{3}{4}$	0.75	75%
$\frac{1}{3}$	0.333333...	$33\frac{1}{3}\%$
$\frac{2}{3}$	0.666666...	$66\frac{2}{3}\%$
$\frac{1}{10}$	0.1	10%
$\frac{2}{10}$	0.2	20%
$\frac{1}{5}$	0.2	20%
$\frac{2}{5}$	0.4	40%

0.3333... and 0.6666... are known as 'recurring' decimals — the same pattern of numbers carries on repeating itself forever. See p20.

The more of those conversions you learn, the better — but for those that you <u>don't know</u>, you must <u>also learn</u> how to <u>convert</u> between the three types. These are the methods:

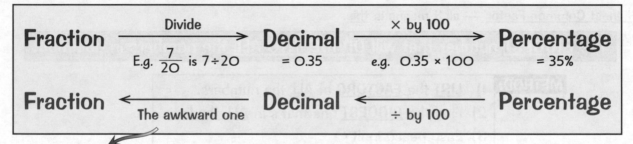

$$\textbf{Fraction} \xrightarrow{\text{Divide}} \textbf{Decimal} \xrightarrow{\times \text{ by 100}} \textbf{Percentage}$$

E.g. $\frac{7}{20}$ is $7 \div 20$ = 0.35   e.g. 0.35 × 100   = 35%

$$\textbf{Fraction} \xleftarrow[\text{The awkward one}]{} \textbf{Decimal} \xleftarrow[\div \text{ by 100}]{} \textbf{Percentage}$$

<u>Converting decimals to fractions</u> is a bit more awkward.
The digits after the decimal point go on the top, and a <u>power of 10</u> on the bottom — with the same number of zeros as there were decimal places.

$$0.6 = \frac{6}{10} \qquad 0.3 = \frac{3}{10} \qquad 0.7 = \frac{7}{10} \quad \text{etc.}$$

$$0.12 = \frac{12}{100} \qquad 0.78 = \frac{78}{100} \qquad 0.05 = \frac{5}{100} \quad \text{etc.}$$

$$0.345 = \frac{345}{1000} \qquad 0.908 = \frac{908}{1000} \qquad 0.024 = \frac{24}{1000} \quad \text{etc.}$$

These can often be <u>cancelled down</u> — see p17.

Eight out of ten cats prefer the perfume Eighty Purr Scent...

Learn the whole of the top table and the 4 conversion processes. Then it's time to break into a mild sweat...

Q1   Turn the following decimals into fractions and reduce them to their simplest form.
        a) 0.4   b) 0.02   c) 0.77   d) 0.555   e) 5.6   [5 marks] **(E)**

Q2   Which is greater: a) 57% or $\frac{5}{9}$, b) 0.2 or $\frac{6}{25}$, c) $\frac{7}{8}$ or 90%?   [3 marks] **(E)**

Fractions

These pages show you how to cope with fraction calculations without your beloved calculator.

Equivalent Fractions and Cancelling Down

1) Equivalent fractions are fractions that are equal in value, even though they look different.

2) You can make equivalent fractions by simply MULTIPLYING (or dividing) top and bottom by the SAME NUMBER each time.

3) To cancel down or simplify a fraction, divide top and bottom by the same number. 'Simplify as far as possible', 'write the fraction in its simplest form' and 'write the fraction in its lowest terms' all mean 'keep dividing till they won't go any further'.

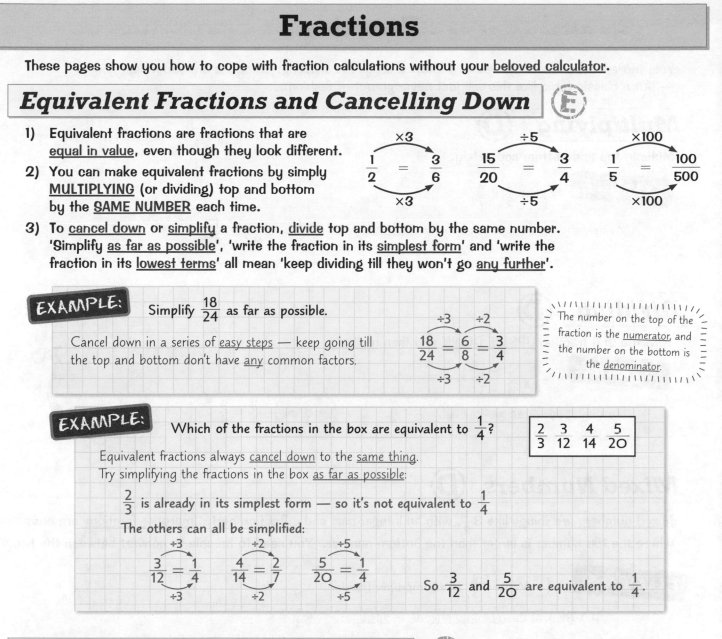

EXAMPLE: Simplify $\frac{18}{24}$ as far as possible.

Cancel down in a series of easy steps — keep going till the top and bottom don't have any common factors.

$$\frac{18}{24} = \frac{6}{8} = \frac{3}{4}$$

The number on the top of the fraction is the numerator, and the number on the bottom is the denominator.

EXAMPLE: Which of the fractions in the box are equivalent to $\frac{1}{4}$?

$$\frac{2}{3} \quad \frac{3}{12} \quad \frac{4}{14} \quad \frac{5}{20}$$

Equivalent fractions always cancel down to the same thing.
Try simplifying the fractions in the box as far as possible:

$\frac{2}{3}$ is already in its simplest form — so it's not equivalent to $\frac{1}{4}$

The others can all be simplified:

$$\frac{3}{12} = \frac{1}{4} \qquad \frac{4}{14} = \frac{2}{7} \qquad \frac{5}{20} = \frac{1}{4}$$

So $\frac{3}{12}$ and $\frac{5}{20}$ are equivalent to $\frac{1}{4}$.

Finding a Fraction of Something

1) When you're asked to find a fraction of something, you can swap the 'of' for a × sign.

2) To multiply a number by a fraction, multiply it by the TOP of the fraction, and divide it by the BOTTOM. It doesn't matter which order you do those two steps in — just start with whatever's easiest.

EXAMPLE: What is $\frac{9}{20}$ of £360?

1) Swap the 'of' for a 'x'.
2) Now you need to multiply by 9 and divide by 20. The division's easier, so start with that.

$$\frac{9}{20} \text{ of £360} = \frac{9}{20} \times £360$$
$$= (£360 \div 20) \times 9$$
$$= £18 \times 9$$
$$= £162$$

LIGHTS... CAMERA... fffrrrACTION...

So many fractions questions end with 'give your answer in its simplest form'. Make sure you can do that, and it's bound to net you a mark or two come exam time. Now try these Exam Practice Questions:

Q1 Which fraction from the list $\frac{50}{75}$, $\frac{10}{15}$, $\frac{8}{12}$, $\frac{4}{9}$ is not equivalent to $\frac{2}{3}$? [1 mark] (E)

Q2 What is $\frac{3}{5}$ of 150 kg? [2 marks] (E)

Fractions

Yes, more fractions stuff to learn. I know what you're thinking: 'fractions are soooo last year' — but a classic topic like this will just never go out of fashion...

Multiplying (D)

Multiply top and bottom separately.

> Remember that multiplying by e.g. $\frac{1}{4}$ is the same as dividing by 4.

EXAMPLE: Find $\frac{3}{5} \times \frac{4}{7}$.

Multiply the top and bottom numbers <u>separately</u>: $\frac{3}{5} \times \frac{4}{7} = \frac{3 \times 4}{5 \times 7} = \frac{12}{35}$

Dividing (D)

Turn the 2nd fraction <u>UPSIDE DOWN</u> and then <u>multiply</u>:

EXAMPLE: Find $\frac{3}{4} \div \frac{1}{3}$.

Turn $\frac{1}{3}$ <u>upside down</u> and <u>multiply</u>: $\frac{3}{4} \div \frac{1}{3} = \frac{3}{4} \times \frac{3}{1} = \frac{3 \times 3}{4 \times 1} = \frac{9}{4}$

Mixed Numbers (D)

<u>Mixed numbers</u> are things like $3\frac{1}{3}$, with an integer part and a fraction part. <u>Improper fractions</u> are ones where the top number is larger than the bottom number. You need to be able to convert between the two.

EXAMPLES:

1. Write $4\frac{2}{3}$ as an improper fraction.

1) Think of the <u>mixed number</u> as an <u>addition</u>: $4\frac{2}{3} = 4 + \frac{2}{3}$

2) Turn the <u>whole number part</u> into a <u>fraction</u>: $4 + \frac{2}{3} = \frac{12}{3} + \frac{2}{3} = \frac{12 + 2}{3} = \frac{14}{3}$

2. Write $\frac{31}{4}$ as a mixed number.

<u>Divide</u> the top number by the bottom.
1) The <u>answer</u> gives the <u>whole number part</u>.
2) The <u>remainder</u> goes <u>on top</u> of the fraction.

$31 \div 4 = 7$ remainder 3
so $\frac{31}{4} = 7\frac{3}{4}$

> If you have to do a calculation with mixed numbers, just turn them into improper fractions first, then carry on as normal. (You might have to change the answer back to a mixed number at the end.)

No fractions were harmed in the making of these pages...

...although one was slightly frightened for a while, and several were tickled.

When you think you've learnt all this, try all of these Exam Practice Questions without a calculator.

Q1   Find: a) $\frac{4}{5} \times \frac{3}{7}$ b) $\frac{9}{10} \div \frac{3}{5}$. Give your answers in their simplest form.   [2 marks] (D)

Q2   a) Write $4\frac{1}{9}$ as an improper fraction.   b) Write $\frac{17}{3}$ as a mixed number.   [2 marks] (D)

Q3   Calculate: a) $\frac{3}{8} \times 1\frac{5}{12}$ b) $1\frac{7}{9} \div 2\frac{2}{3}$. Give your answers in their simplest form. [6 marks] (C)

Fractions

There's an awkward-looking maths word looming large on this page — but don't let it put you off. Remember, the denominator is just the number on the bottom of the fraction.

Common Denominators (D)

This comes in handy for <u>comparing</u> the sizes of fractions and for <u>adding</u> or <u>subtracting</u> fractions.

> You need to find a number that <u>all</u> the denominators <u>divide into</u> — this will be your <u>common denominator</u>. The simplest way is to <u>multiply</u> all the different denominators together.

EXAMPLE: Put these fractions in ascending order of size: $\frac{8}{3}, \frac{5}{4}, \frac{12}{5}$

1) The <u>new denominator</u> has to be a number all the denominators <u>divide into</u>:

3, 4 and 5 all go into $3 \times 4 \times 5 = 60$. Make 60 the common denominator.

2) Then <u>change each fraction</u> so it's over the <u>new number</u>:

$\frac{8}{3} = \frac{160}{60}$ (×20) $\frac{5}{4} = \frac{75}{60}$ (×15) $\frac{12}{5} = \frac{144}{60}$ (×12)

3) Now they're easy to <u>write in order</u>:

So the correct order is
$\frac{75}{60}, \frac{144}{60}, \frac{160}{60}$, or $\frac{5}{4}, \frac{12}{5}, \frac{8}{3}$

Use the <u>original</u> <u>fractions</u> in the final answer.

To Add and Subtract — sort the denominators first (D)

> 1) Make sure the denominators are <u>the same</u> (see above).
>
> 2) Add (or subtract) the top lines (numerators) <u>only</u>.

If you're adding or subtracting <u>mixed numbers</u>, it usually helps to convert them to improper fractions first.

EXAMPLES:

1. Calculate $\frac{1}{2} - \frac{1}{5}$.

Find a <u>common denominator</u>: $\frac{1}{2} - \frac{1}{5} = \frac{5}{10} - \frac{2}{10}$

Combine the <u>top lines</u>: $= \frac{5-2}{10} = \frac{3}{10}$

2. Work out $2\frac{4}{7} + 1\frac{5}{7}$.

1) Write the mixed numbers as <u>improper fractions</u>: $2\frac{4}{7} + 1\frac{5}{7} = \frac{18}{7} + \frac{12}{7}$

2) The <u>denominators</u> are the same, so just add the <u>top lines</u>: $= \frac{18+12}{7}$

3) Turn the answer back into a <u>mixed number</u>: $= \frac{30}{7} = 4\frac{2}{7}$

Oh, how beastly — these denominators are so common...

Please, please don't try to add or subtract fractions with different numbers on the bottom. You'll just get into a terrible pickle if you do. Try these questions to see if you've cracked the stuff on this page.

Q1 Put the following fractions in order from smallest to largest: $\frac{3}{4}, \frac{11}{20}, \frac{7}{10}, \frac{5}{8}$ [5 marks] (D)

Q2 Find the following, giving your answers in their simplest form:

a) $\frac{4}{15} + \frac{1}{3}$ b) $\frac{19}{24} - \frac{5}{8}$ [4 marks] (D)

Fractions and Recurring Decimals

You might think that a decimal is just a decimal. But oh no — things get a lot more juicy than that...

Recurring or Terminating... ©

1) **Recurring** decimals have a **pattern** of numbers which repeats forever.

   For example, $\frac{1}{3}$ is the decimal 0.333333...

2) It doesn't have to be a single digit that repeats.

   E.g. You could have 0.143143143...

3) The **repeating part** is usually marked with **dots** on top of the number.

4) If there's **one dot**, only **one digit** is repeated. If there are **two dots**, then **everything from the first dot to the second dot** is the repeating bit.

   E.g. $0.2\dot{5} = 0.2555555...$,
   $0.\dot{2}\dot{5} = 0.25252525...$,
   $0.\dot{2}6\dot{5} = 0.265265265...$

5) **Terminating** decimals **don't** go on forever.

   E.g. $\frac{1}{20}$ is the terminating decimal 0.05

6) **All** terminating and recurring decimals can be written as **fractions**.

Fraction	Recurring decimal or terminating decimal?	Decimal
$\frac{1}{2}$	Terminating	0.5
$\frac{1}{3}$	Recurring	$0.\dot{3}$
$\frac{1}{4}$	Terminating	0.25
$\frac{1}{5}$	Terminating	0.2
$\frac{1}{6}$	Recurring	$0.1\dot{6}$
$\frac{1}{7}$	Recurring	$0.\dot{1}4285\dot{7}$
$\frac{1}{8}$	Terminating	0.125
$\frac{1}{9}$	Recurring	$0.\dot{1}$
$\frac{1}{10}$	Terminating	0.1

Turning Fractions into Recurring Decimals ©

You might find this cropping up in your exam too — and if they're being really unpleasant, they'll stick it in a **non-calculator** paper.

EXAMPLE: Without using a calculator, write $\frac{5}{11}$ as a recurring decimal.

1) Remember, $\frac{5}{11}$ means $5 \div 11$, so you can just _do the division_. The trick is to treat the 5 as a decimal — write it as 5.000...

   For more about division, see p9-10.

   11 into 50 goes 4 times...
   $$\begin{array}{r} 0.4 \\ 11\overline{)5.5^06\,0\,0\,0} \end{array}$$
   ...and carry the 6

   11 into 60 goes 5 times...
   $$\begin{array}{r} 0.4\ 5 \\ 11\overline{)5.5^06^05\,0\,0} \end{array}$$
   ...and carry the 5

   $$\begin{array}{r} 0.4\ 5\ 4\ 5 \\ 11\overline{)5.5^06^05^06^05^0} \end{array}$$

2) Keep going until you can see the _repeating pattern_. Write the recurring decimal using dots above the repeating part.

   $5 \div 11 = 0.454545...$

   so $\frac{5}{11} = 0.\dot{4}\dot{5}$

Oh, what's recurrin'?...

This seems pretty tricky, I admit, but you'll be on the right track if you know what those dots on top of a decimal mean, and how to turn a fraction into a decimal by dividing — even without a calculator.

Q1   Without a calculator, use division to show that $\frac{1}{6} = 0.1\dot{6}$ [3 marks] ©

Proportion Problems

Proportion problems all involve amounts that increase or decrease together. Awww.

Learn the Golden Rule for Proportion Questions (D)

There are lots of exam questions which at first sight seem completely
different but in fact they can all be done using the GOLDEN RULE...

DIVIDE FOR ONE, THEN TIMES FOR ALL

EXAMPLE: 5 pints of milk cost £1.30. How much will 3 pints cost?

The GOLDEN RULE says: **DIVIDE FOR ONE, THEN TIMES FOR ALL**

which means: Divide the price by 5 to find how much FOR ONE PINT,
then multiply by 3 to find how much FOR 3 PINTS.

So for 1 pint: £1.30 ÷ 5 = 0.26 = 26p
For 3 pints: 26p × 3 = 78p

My favourite cereal is muesli.

Use the Golden Rule to Scale Recipes Up or Down (D)

EXAMPLE: Judy is making fruit punch using the recipe
shown on the right. She wants to make
enough to serve 20 people.
How much of each ingredient will Judy need?

Fruit Punch (serves 8)

800 ml orange juice
600 ml grape juice
200 ml cherry juice
140 g fresh pineapple

Use the GOLDEN RULE again: **DIVIDE FOR ONE, THEN TIMES FOR ALL**

which means: Divide each amount by 8 to find how much FOR ONE PERSON,
then multiply by 20 to find how much FOR 20 PEOPLE.

So for 1 person you need: And for 20 people you need:

800 ml ÷ 8 = 100 ml orange juice ⇒ 20 × 100 ml = 2000 ml orange juice

600 ml ÷ 8 = 75 ml grape juice   ⇒ 20 × 75 ml = 1500 ml grape juice

200 ml ÷ 8 = 25 ml cherry juice ⇒ 20 × 25 ml = 500 ml cherry juice

140 g ÷ 8 = 17.5 g pineapple   ⇒ 20 × 17.5 g = 350 g pineapple

For some questions like this, you can just multiply — e.g. in the example above, if you wanted to know the
ingredients for 16 servings of punch, you could just times everything in the recipe by 2.

This is OK if you're confident you know what you're doing, but remember, the GOLDEN RULE always works...

The Three Mathsketeers say "divide for one, then times for all"...

It's a simple rule — the trick is knowing when to use it. Learning the examples above will help.

Q1   If seven pencils cost 98p, how much will 4 pencils cost?   [3 marks] (D)

Q2   To make 4 servings of leek and liquorice soup, you need 600 g of leeks and 80 g of liquorice.
       a) What quantity of leeks do you need for 6 servings of soup?   [3 marks] (D)
       b) How much liquorice do you need for 9 servings?   [3 marks]

Proportion Problems

Another page, another golden rule — this one's been specially handcrafted to deal with 'best buy' questions.

Best Buy Questions — Find the Amount per Penny (D)

A slightly different type of proportion question is comparing the 'value for money' of 2 or 3 similar items. For these, follow the second <u>GOLDEN RULE</u>...

Divide by the <u>PRICE</u> in pence (to get the amount <u>per penny</u>)

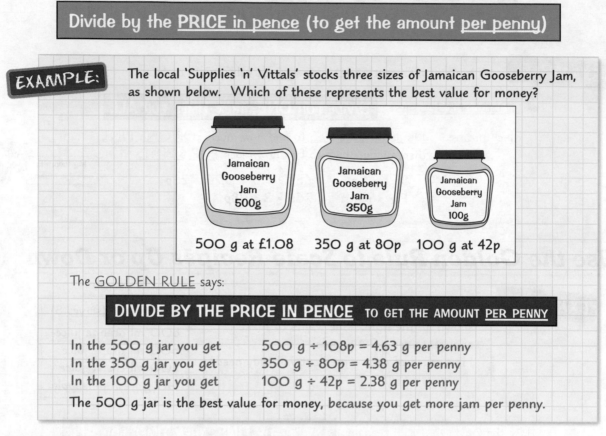

EXAMPLE: The local 'Supplies 'n' Vittals' stocks three sizes of Jamaican Gooseberry Jam, as shown below. Which of these represents the best value for money?

500 g at £1.08   350 g at 80p   100 g at 42p

The <u>GOLDEN RULE</u> says:

DIVIDE BY THE PRICE <u>IN PENCE</u> TO GET THE AMOUNT <u>PER PENNY</u>

In the 500 g jar you get	500 g ÷ 108p = 4.63 g per penny
In the 350 g jar you get	350 g ÷ 80p = 4.38 g per penny
In the 100 g jar you get	100 g ÷ 42p = 2.38 g per penny

The 500 g jar is the best value for money, because you get more jam per penny.

With any question comparing 'value for money', <u>DIVIDE BY THE PRICE</u> (in pence) and it will always be the <u>BIGGEST ANSWER</u> that is the <u>BEST VALUE FOR MONEY</u>.

...or Find the Price per Unit (D)

For some questions, the numbers mean it's easier to <u>divide by the amount</u> to get the <u>cost per unit</u> (e.g per gram, per litre, etc.). In that case, the <u>best buy</u> is the <u>smallest answer</u> — the <u>lowest cost</u> per unit. Doing the example above in this way, you'd get:

The jam in the 500 g jar costs	108p ÷ 500 g = 0.216p per gram
The jam in the 350 g jar costs	80p ÷ 350 g = 0.229p per gram
The jam in the 100 g jar costs	42p ÷ 100 g = 0.42p per gram

The 500 g jar is the best value for money, because it's the cheapest per gram.

This page is the best by far...

Don't forget what you're looking for in a best buy question — the best buy gives you the biggest amount per penny, and has the lowest cost per unit. Try this Exam Practice Question:

Q1   Tomato ketchup comes in bottles of three sizes:
342 g for 93p, 550 g for £1.48, 910 g for £2.79
List the bottles in order of value for money, from best to worst.   [3 marks] (D)

Percentages

There are lots of different types of percentage questions. Read the examples on the next two pages carefully and make sure you can recognise the different percentage questions you might meet.

Three Simple Question Types (D)

Type 1 — "Find x% of y"

Turn the percentage into a <u>decimal</u>, then <u>multiply</u>.

EXAMPLE: Find 15% of £46.

divide by 100 to turn a percentage into a decimal

1) Write 15% as a <u>decimal</u>: $15\% = 15 \div 100 = 0.15$
2) <u>Multiply</u> £46 by 0.15: $0.15 \times £46 = £6.90$

Type 2 — "Find the new amount after a % increase/decrease"

1) Work out the "<u>% of original value</u>" as above — this is the actual increase or decrease.
2) <u>Add to or subtract from</u> the original value.

EXAMPLE: A toaster is reduced in price by 40% in the sales. It originally cost £68. What is the new price of the toaster?

1) Find <u>40% of £68</u> (using the method above): $40\% = 40 \div 100 = 0.4$
So 40% of £68 = $0.4 \times £68 = £27.20$
2) It's a <u>decrease</u>, so <u>subtract</u> from the original: So the new price is: £68 − £27.20 = £40.80

Or if you prefer, you can use the <u>multiplier</u> method:

1) Write 40% as a <u>decimal</u>: $40\% = 40 \div 100 = 0.4$
2) It's a decrease, so find the <u>multiplier</u> by taking 0.4 from 1: (For an increase, you'd <u>add it to</u> 1 instead.) multiplier = 1 − 0.4 = 0.6
3) Multiply the <u>original</u> by the <u>multiplier</u>: $£68 \times 0.6 = £40.80$

Type 3 — "Express x as a percentage of y"

<u>Divide</u> x by y, then multiply by <u>100</u>.

EXAMPLE: Give 40p as a percentage of £3.34.

1) Make sure both amounts are in the <u>same units</u> — convert £3.34 to pence: £3.34 = 334p
2) <u>Divide</u> 40p by 334p, <u>then multiply</u> by 100: $(40 \div 334) \times 100 = 12.0\%$ (to 1 d.p.)

Fact: 70% of people understand percentages, the other 40% don't...

Learn the details for each type of percentage question, then turn over and write it all down.
Then try these Exam Practice Questions:

Q1 A normal bottle of Kenny's Kiwi Juice contains 450 ml of juice. A special offer bottle contains 22% extra. How much juice is in the special offer bottle? [2 marks] (D)

Q2 Jenny and Penny bought a llama together for £4500. Jenny paid £1215 towards this total. What percentage of the cost of the llama did Jenny pay? [3 marks] (D)

Percentages

That's right, there are more percentage questions to learn over here. Sorry about that.

Percentages Without a Calculator (D)

Don't worry if you get a <u>non-calculator</u> percentage question.
You can use the trusty rules for <u>dividing by 10 and 100</u> (see p7) to help you work things out.

EXAMPLE: Calculate 23% of 250 g. Show your working.

1) You know that <u>250 g is 100%</u>, so it's easy to find <u>10%</u> and <u>1%</u>:

$$100\% = 250 \text{ g}$$
$$\div 10 \searrow 10\% = 25 \text{ g} \qquad \div 10$$
$$\div 10 \searrow 1\% = 2.5 \text{ g} \qquad \div 10$$

2) Now use those values to <u>make 23%</u>:

$$20\% = 2 \times 10\%$$
$$= 2 \times 25 \text{ g} = 50 \text{ g}$$
$$3\% = 3 \times 1\%$$
$$= 3 \times 2.5 \text{ g} = 7.5 \text{ g}$$
$$\text{So } 23\% = 20\% + 3\%$$
$$= 50 \text{ g} + 7.5 \text{ g}$$
$$= 57.5 \text{ g}$$

Simple Interest (D)

1) <u>Interest</u> is money that's usually paid when you borrow or save some money. It gets added to the <u>original amount</u> you saved or borrowed.

2) <u>Simple interest</u> means a certain percentage of the <u>original amount</u> is paid at regular intervals (usually once a year). So the amount of interest is <u>the same every time</u> it's paid.

3) To work out the <u>total amount</u> of simple interest paid over a certain amount of time, just find the size of <u>one interest payment</u>, then multiply it by the <u>number of payments</u> in the time period.

EXAMPLE:

Regina invests £380 in an account which pays 3% simple interest per annum. How much interest will she earn in 4 years?

'Per annum' just means 'each year'.

1) Work out the amount of interest earned <u>in one year</u>:

$$3\% = 3 \div 100 = 0.03$$
$$3\% \text{ of } £380 = 0.03 \times £380$$
$$= £11.40$$

2) Multiply by 4 to get the <u>total interest</u> for <u>4 years</u>:

$$4 \times £11.40 = £45.60$$

Simple interest — it's simple, but it's not that interesting...

Two special types of percentage questions here — make sure you get to grips with both. Trying to work out a percentage without a calculator sounds nasty at first, but with some easy dividing and a bit of adding, you can do it surprisingly easily. Now have a go at these:

Q1 Without using a calculator, find 45% of 180 kg. Show your working. [2 marks] (D)

Q2 Benny invests £1900 for 5 years in an account which pays simple interest at a rate of 2.2% per annum. How much interest will Benny earn in total? [3 marks] (D)

Ratios

Ratios can be a murky topic — but work through these examples, and it should all become crystal clear...

Reducing Ratios to their Simplest Form Ⓓ

To reduce a ratio to a <u>simpler form</u>, divide <u>all the numbers</u> in the ratio by the <u>same thing</u> (a bit like simplifying a fraction). It's in its <u>simplest form</u> when there's nothing left you can divide by.

> **EXAMPLE:** Write the ratio 15:18 in its simplest form.
>
> For the ratio 15:18, both numbers have a <u>factor</u> of 3, so <u>divide them by</u> 3.
>
> $$\div 3 \left(\begin{array}{c} 15:18 \\ = \quad 5:6 \end{array} \right) \div 3$$
>
> You can't reduce this any further. So the simplest form of 15:18 is **5:6**.

A handy trick for the calculator paper — use the fraction button

If you enter a fraction with the 🔲 or 🔲 button, the calculator will cancel it down when you press ⏹.

So for 8:12, enter $\frac{8}{12}$ as a fraction and it'll get reduced to $\frac{2}{3}$. Now just change it back to a ratio, i.e. <u>**2 : 3**</u>. Ace.

Proportional Division Ⓓ

If you know the <u>ratio</u> and the <u>TOTAL AMOUNT</u> you can split the total into <u>separate amounts</u>.

The key is to think about the <u>PARTS</u> that make up each amount — just follow these three steps:

> 1) <u>ADD UP THE PARTS</u>
> 2) <u>DIVIDE TO FIND ONE "PART"</u>
> 3) <u>MULTIPLY TO FIND THE AMOUNTS</u>

> **EXAMPLE:** Jess, Mo and Greg share £9100 in the ratio 2:4:7. How much does Mo get?
>
> 1) The ratio 2:4:7 means there will be a total of 13 <u>parts</u>:   2 + 4 + 7 = 13 parts
> 2) Divide the <u>total amount</u> by the number of <u>parts</u>:   £9100 ÷ 13 = £700   (= 1 part)
> 3) We want to know <u>Mo's share</u>, which is <u>4 parts</u>:   4 parts = 4 × 700 = £2800

Scaling Up Ratios Ⓓ

If you know the <u>ratio</u> and the actual size of <u>ONE AMOUNT</u>, you can <u>scale the ratio up</u> to find the other amounts.

> **EXAMPLE:** Mortar is made from sand and cement in the ratio 7:2. If 21 buckets of sand are used, how much cement is needed?
>
> You need to <u>multiply by 3</u> to go from 7 to 21 on the left-hand side — do that to <u>both sides</u>:
>
>
>
> sand:cement
> $$= \times 3 \left(\begin{array}{c} 7:2 \\ 21:6 \end{array} \right) \times 3$$
>
> So **6 buckets** of cement are needed.

Ratio Nelson — he proportionally divided the French at Trafalgar...

Learn the rules for simplifying, the three steps for proportional division and how to scale ratios up. Now turn over and write down what you've learned. Then try these:

Q1   I have 28 blue bow ties and 14 red bow ties.
      Write the ratio of blue to red bow ties in my collection in its simplest form.   [2 marks] Ⓓ

Q2   Divide £8400 in the ratio 5:3:4.   [3 marks] Ⓓ

Rounding Off

You need to be able to use <u>3 different rounding methods</u>.
We'll do decimal places first, but there's the same basic idea behind all three.

Decimal Places (d.p.)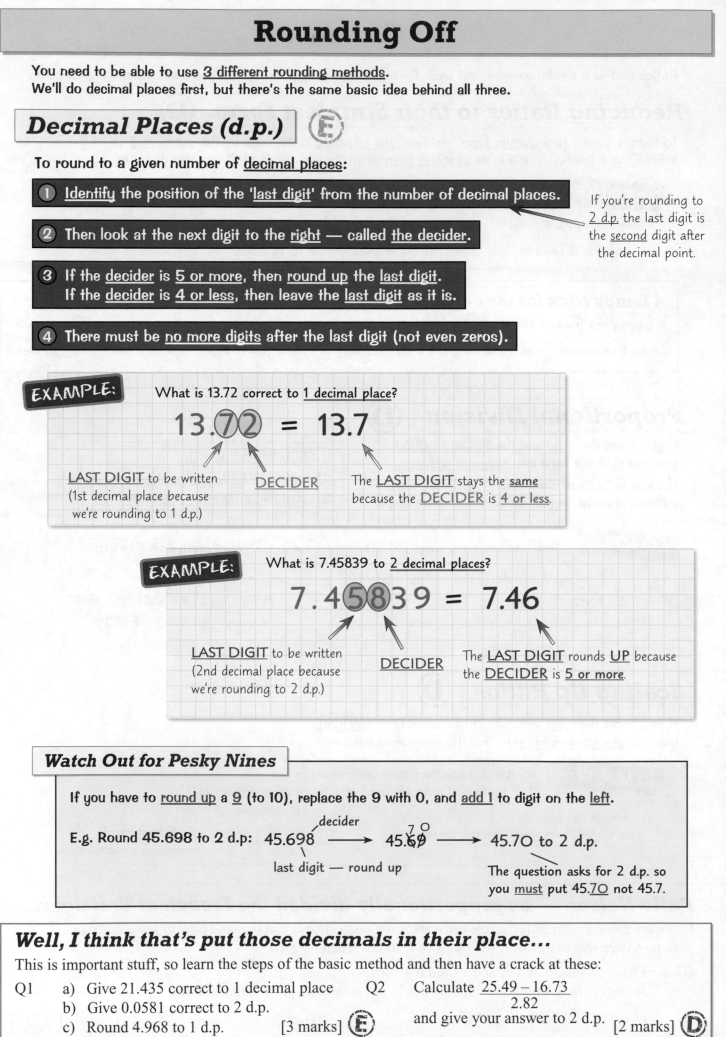

To round to a given number of <u>decimal places</u>:

① <u>Identify</u> the position of the '<u>last digit</u>' from the number of decimal places.

② Then look at the next digit to the <u>right</u> — called <u>the decider</u>.

③ If the <u>decider</u> is <u>5 or more</u>, then <u>round up</u> the <u>last digit</u>.
If the <u>decider</u> is <u>4 or less</u>, then leave the <u>last digit</u> as it is.

④ There must be <u>no more digits</u> after the last digit (not even zeros).

> If you're rounding to <u>2 d.p.</u> the last digit is the <u>second</u> digit after the decimal point.

EXAMPLE: What is 13.72 correct to <u>1 decimal place</u>?

$$13.\underline{72} = 13.7$$

<u>LAST DIGIT</u> to be written (1st decimal place because we're rounding to 1 d.p.)

<u>DECIDER</u>

The <u>LAST DIGIT</u> stays the <u>same</u> because the <u>DECIDER</u> is <u>4 or less</u>.

EXAMPLE: What is 7.45839 to <u>2 decimal places</u>?

$$7.4\underline{58}39 = 7.46$$

<u>LAST DIGIT</u> to be written (2nd decimal place because we're rounding to 2 d.p.)

<u>DECIDER</u>

The <u>LAST DIGIT</u> rounds <u>UP</u> because the <u>DECIDER</u> is <u>5 or more</u>.

Watch Out for Pesky Nines

If you have to <u>round up</u> a <u>9</u> (to 10), replace the 9 with 0, and <u>add 1</u> to digit on the <u>left</u>.

E.g. Round **45.698 to 2 d.p:** 45.698 $\longrightarrow$ 45.6̶9̶⁷⁰ $\longrightarrow$ 45.70 to 2 d.p.

decider

last digit — round up

The question asks for 2 d.p. so you <u>must</u> put 45.<u>70</u> not 45.7.

Well, I think that's put those decimals in their place...

This is important stuff, so learn the steps of the basic method and then have a crack at these:

Q1   a) Give 21.435 correct to 1 decimal place
       b) Give 0.0581 correct to 2 d.p.
       c) Round 4.968 to 1 d.p.   [3 marks] **(E)**

Q2   Calculate $\dfrac{25.49 - 16.73}{2.82}$
and give your answer to 2 d.p.   [2 marks] **(D)**

Rounding Off

Significant Figures (s.f.) 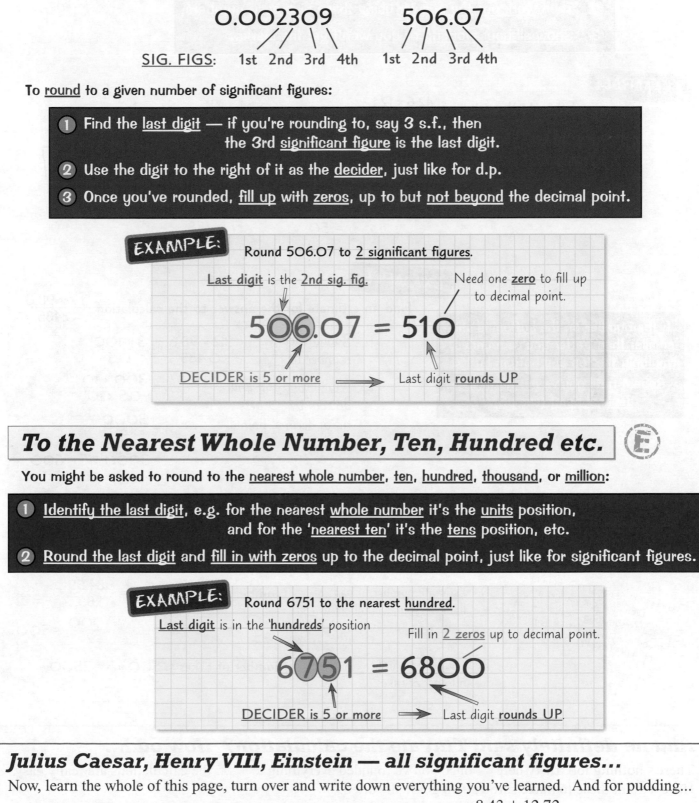 Ⓓ

The <u>1st significant figure</u> of any number is <u>the first digit which isn't a zero</u>.

The <u>2nd, 3rd, 4th, etc. significant figures</u> follow immediately after the 1st — they're allowed to be zeros.

$$0.002309 \qquad 506.07$$

SIG. FIGS: 1st   2nd   3rd   4th   1st   2nd   3rd   4th

To <u>round</u> to a given number of significant figures:

① Find the <u>last digit</u> — if you're rounding to, say 3 s.f., then the 3rd <u>significant figure</u> is the last digit.

② Use the digit to the right of it as the <u>decider</u>, just like for d.p.

③ Once you've rounded, <u>fill up</u> with <u>zeros</u>, up to but <u>not beyond</u> the decimal point.

EXAMPLE: Round 506.07 to <u>2 significant figures</u>.

<u>Last digit</u> is the <u>2nd sig. fig.</u>   Need one <u>zero</u> to fill up to decimal point.

$$506.07 = 510$$

<u>DECIDER is 5 or more</u> ⟶ Last digit <u>rounds UP</u>

To the Nearest Whole Number, Ten, Hundred etc. Ⓔ

You might be asked to round to the <u>nearest whole number</u>, <u>ten</u>, <u>hundred</u>, <u>thousand</u>, or <u>million</u>:

① <u>Identify the last digit</u>, e.g. for the nearest <u>whole number</u> it's the <u>units</u> position, and for the '<u>nearest ten</u>' it's the <u>tens</u> position, etc.

② <u>Round the last digit</u> and <u>fill in with zeros</u> up to the decimal point, just like for significant figures.

EXAMPLE: Round 6751 to the nearest <u>hundred</u>.

<u>Last digit</u> is in the '<u>hundreds</u>' position   Fill in <u>2 zeros</u> up to decimal point.

$$6751 = 6800$$

<u>DECIDER is 5 or more</u> ⟶ Last digit <u>rounds UP</u>.

Julius Caesar, Henry VIII, Einstein — all significant figures...

Now, learn the whole of this page, turn over and write down everything you've learned. And for pudding...

Q1   a) Round 653 to 1 s.f.   b) Round 14.6 to 2 s.f.
     c) Give 168.7 to the nearest whole number.
     d) Give 82 430 to the nearest thousand. [4 marks] Ⓓ

Q2   Calculate $\dfrac{8.43 + 12.72}{5.63 - 1.21}$

and give your answer to 2 s.f.   [2 marks] Ⓓ

Estimating Calculations

"Estimate" doesn't mean "take a wild guess", so don't just make something up...

Estimating ©

Have a look at the previous page to remind yourself how to round to 1 s.f.

1) <u>Round everything off</u> to <u>1 significant figure</u>.

2) Then <u>work out the answer</u> using these nice easy numbers.

3) <u>Show all your working</u> or you won't get the marks.

EXAMPLE:

Estimate the value of $\dfrac{42.6 \times 12.1}{7.9}$.

1) <u>Round</u> each number to <u>1 s.f.</u>

$$\frac{42.6 \times 12.1}{7.9} \approx \frac{40 \times 10}{8}$$

2) Do the <u>calculation</u> with the rounded numbers.

$$= \frac{400}{8}$$

$$= 50$$

≈ means 'approximately equal to'.

EXAMPLE:

Find an estimate for the answer to the calculation $\dfrac{3.2 \times 98.6}{0.485}$.

1) <u>Round</u> each number to <u>1 s.f.</u>

$$\frac{3.2 \times 98.6}{0.485} \approx \frac{3 \times 100}{0.5}$$

2) <u>Multiplying</u> top and bottom by <u>10</u> gets rid of decimal point.

$$= \frac{300 \times 10}{0.5 \times 10}$$

$$= \frac{3000}{5}$$

$$= 3000 \div 5 = 600$$

If the number on the <u>bottom</u> is still <u>smaller than 1</u> after you've rounded, <u>multiply the top and bottom by 10</u>.
Repeat until you've <u>got rid of the decimal point</u> (see p6).

EXAMPLE:

Jo has a cake-making business. She spent <u>£984.69</u> on flour last year.
A bag of flour costs <u>£1.89</u>, and she makes an average of <u>5 cakes from each bag</u> of flour.
Work out an estimate of how many cakes she made last year.

Don't panic if you get a 'real-life' estimating question — just round everything to 1 s.f. as before.

1) Estimate number of bags of flour — <u>round</u> numbers to <u>1 s.f.</u>

Number of bags of flour $= \dfrac{984.69}{1.89}$

$$\approx \frac{1000}{2} = 500$$

2) Multiply to find the number of cakes.

Number of cakes $\approx 500 \times 5 = 2500$

And he definitely said Tim ate the calculation? How odd...

There's nothing too tricky here — once you've rounded everything to 1 s.f. the calculations are really easy. Have a go at these to make sure you've got it all sorted.

Q1 a) Estimate the value of $\dfrac{586.7}{9.8 \times 3.1}$.   b) Estimate the value of $\dfrac{22.3 \times 11.4}{0.532}$.   [4 marks] ©

Q2   Kate buys 17 kg of turnips at a cost of £1.93 per kg. Estimate the total cost.   [2 marks] ©

Square Roots and Cube Roots

Take a deep breath, and get ready to tackle this page. Good luck with it, I'll be rootin' for ya...

Square Roots (F)

'Squared' means 'multiplied by itself': $8^2 = 8 \times 8 = 64$

SQUARE ROOT $\sqrt{}$ is the reverse process: $\sqrt{64} = 8$

The best way to think of it is: **'Square Root' means 'What Number Times by Itself gives...'**

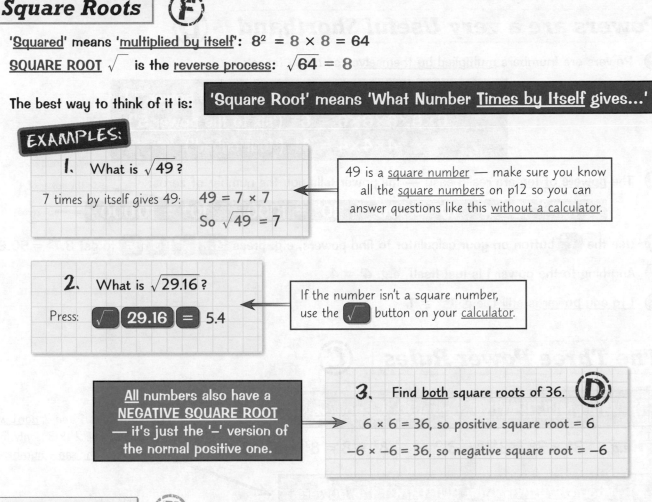

EXAMPLES:

1. What is $\sqrt{49}$?

7 times by itself gives 49: $49 = 7 \times 7$

So $\sqrt{49} = 7$

49 is a square number — make sure you know all the square numbers on p12 so you can answer questions like this without a calculator.

2. What is $\sqrt{29.16}$?

Press: $\sqrt{}$ 29.16 $=$ 5.4

If the number isn't a square number, use the $\sqrt{}$ button on your calculator.

All numbers also have a **NEGATIVE SQUARE ROOT** — it's just the '–' version of the normal positive one.

3. Find both square roots of 36. (D)

$6 \times 6 = 36$, so positive square root = 6

$-6 \times -6 = 36$, so negative square root = -6

Cube Roots (E)

'Cubed' means 'multiplied by itself and then by itself again': $2^3 = 2 \times 2 \times 2 = 8$

CUBE ROOT $^3\sqrt{}$ is the reverse process: $^3\sqrt{8} = 2$

'Cube Root' means 'What Number Times by Itself and then by Itself Again gives...'

You need to be able to write down the cube roots of the cube numbers given on p12 without a calculator. To find the cube root of any other number you can use your calculator — press $^3\sqrt{}$.

EXAMPLES:

1. What is $^3\sqrt{27}$?

27 is a cube number.

3 times by itself and then by itself again gives 27: $27 = 3 \times 3 \times 3$

So $^3\sqrt{27} = 3$

2. What is $^3\sqrt{4913}$?

Press: $^3\sqrt{}$ 4913 $=$ 17

"Cue brute", that's what I call Charley when I play him at snooker...

Once you've got the meanings of square root and cube root well and truly sorted, have a go at these.

Q1 Find a) $\sqrt{196}$ and b) $\sqrt{64}$ without using a calculator. c) What is $\sqrt{56.25}$? [3 marks] (F)

Q2 Find a) $^3\sqrt{125}$ and b) $^3\sqrt{1000}$ without using a calculator. c) What is $^3\sqrt{9261}$? [3 marks] (E)

Powers

You've already seen 'to the power 2' and 'to the power 3' — they're just 'squared' and 'cubed'. They're just the tip of the iceberg — any number can be a power if it puts its mind to it...

Powers are a very Useful Shorthand (D)

1 Powers are 'numbers <u>multiplied by themselves</u> so many times':

$$2\times2\times2\times2\times2\times2\times2 = 2^7 \text{ ('two to the power 7')}$$
$$6\times6\times6\times6\times6 = 6^5 \text{ ('six to the power 5')}$$
$$4\times4\times4 = 4^3 \text{ ('four cubed')}$$

2 The <u>powers of ten</u> are really easy — the power tells you the number of zeros:

$$10^1 = 10 \qquad 10^2 = 100 \qquad 10^3 = 1000 \qquad 10^4 = 10\,000$$

to the power of 4 • 4 zeros

3 Use the $x^\blacksquare$ button on your calculator to find powers, e.g. press 3.7 $x^\blacksquare$ 3 = to get $3.7^3 = 50.653$.

4 Anything to the <u>power 1</u> is just <u>itself</u>, e.g. $4^1 = 4$.

5 <u>1 to any power</u> is <u>still 1</u>, e.g. $1^{457} = 1$.

The Three Power Rules (C)

1) When <u>MULTIPLYING</u>, you <u>ADD</u> the powers.

e.g. $3^4 \times 3^6 = 3^{4+6} = 3^{10}$ $\qquad 8^3 \times 8 = 8^3 \times 8^1 = 8^{3+1} = 8^4$

<u>Warning</u>: Rules 1 and 2 <u>don't work</u> for things like $2^3 \times 3^7$, only for <u>powers of the same number</u>.

2) When <u>DIVIDING</u>, you <u>SUBTRACT</u> the powers.

e.g. $5^4 \div 5^2 = 5^{4-2} = 5^2$ $\qquad p^8 \div p^7 = p^{8-7} = p^1 = p$

Don't be put off by <u>letters</u> — they obey the <u>same rules</u>.

3) When <u>RAISING</u> one power to another, you <u>MULTIPLY</u> the powers.

e.g. $(4^2)^4 = 4^{2\times4} = 4^8$, $\quad (x^4)^6 = x^{4\times6} = x^{24}$

EXAMPLE: $a = 5^9$ and $b = 5^4 \times 5^2$. What is the value of $\frac{a}{b}$?

1) Work out b — <u>add</u> the powers: $\qquad b = 5^4 \times 5^2 = 5^{4+2} = 5^6$

2) <u>Divide</u> a by b — <u>subtract</u> the powers: $\quad \frac{a}{b} = 5^9 \div 5^6 = 5^{9-6}$
$$= 5^3 = 125$$

"I've got the power! oh, oh, oh, oh..." Oh no, I'm feeling kinda 90s...

Learn this page off by heart, then cover it up and have a go at these...

Q1 a) Find $3^3 + 4^2$ without a calculator.
b) Use your calculator to find 6.2^3.
c) Write one million as a power of 10.
[4 marks] (D)

Q2 Simplify a) $4^2 \times 4^3$ b) $7^6 \div 7^3$ c) $(q^2)^4$ [3 marks] (C)

Q3 Find $\frac{6^3 \times 6^5}{6^6}$ without using a calculator. [2 marks] (C)

Revision Questions for Section One

Well, that wraps up <u>Section One</u> — time to put yourself to the test and find out <u>how much you really know</u>.
- Try these questions and <u>tick off each one</u> when you <u>get it right</u>.
- When you've done <u>all the questions</u> for a topic and are <u>completely happy</u> with it, tick off the topic.

<u>Ordering Numbers and Arithmetic (p2-10)</u> ☑

1) Write this number out in words: 21 306 515
2) Put these numbers in order of size: 2.2, 4.7, 3.8, 3.91, 2.09, 3.51

<u>Don't</u> use your calculator for questions 3-5
3) Calculate: a) 258 + 624   b) 533 – 87   c) £2.30 + £1.12 + 75p
4) Find: a) £1.20 × 100   b) £150 ÷ 300
5) Work out: a) 51 × 27   b) 338 ÷ 13   c) 3.3 × 19   d) 4.2 ÷ 12

<u>Types of Number, Factors and Multiples (p11-15)</u> ☑

6) Find: a) –10 – 6   b) –35 ÷ –5   c) –4 + –5 + 22 – –7
7) What are square numbers? Write down the first ten of them.
8) Find all the prime numbers between 40 and 60 (there are 5 of them).
9) What are multiples? Find the first six multiples of: a) 10   b) 4
10) Express each of these as a product of prime factors: a) 210   b) 1050
11) Find: a) the HCF of 42 and 28   b) the LCM of 8 and 10

<u>Fractions and Decimals (p16-20)</u> ☑

12) Write: a) 0.04 as: (i) a fraction (ii) a percentage   b) 65% as: (i) a fraction (ii) a decimal
13) How do you simplify a fraction?
14) Calculate a) $\frac{4}{7}$ of 560   b) $\frac{2}{5}$ of £150
15) Work out without a calculator: a) $\frac{5}{8}+\frac{9}{4}$ b) $\frac{2}{3}-\frac{1}{7}$ c) $\frac{25}{6}\div\frac{8}{3}$ d) $\frac{2}{3}\times 4\frac{2}{5}$
16) What is a recurring decimal? How do you show that a decimal is recurring?

<u>Proportions (p21-22)</u> ☑

17) Rick ordered 5 pints of milk from the milkman. His bill was £2.35. How much would 3 pints cost?
18) Tins of Froggatt's Ham come in two sizes. Which is the best buy, 100g for 24p or 250 g for 52p?

<u>Percentages (p23-24)</u> ☑

19) What's the method for finding one amount as a percentage of another?
20) A DVD player costs £50 plus VAT. If VAT is 20%, how much does the DVD player cost?
21) A top that should cost £45 has been reduced by 15%. Carl has £35. Can he afford the top?

<u>Ratios (p25)</u> ☑

22) Sarah has 150 carrots and 240 turnips. Write the ratio of carrots to turnips in its simplest form.
23) Divide 3000 in the ratio 5:8:12.
24) To get her favourite shade of purple, Tia mixes blue and red paint in the ratio 8:3.
How much red paint should she mix with 72 litres of blue paint?

<u>Rounding and Estimating (p26-28)</u> ☑

25) Round a) 17.65 to 1 d.p. b) 6743 to 2 s.f. c) 3 643 510 to the nearest million.
26) Estimate the value of a) $\frac{17.8\times 32.3}{6.4}$ b) $\frac{96.2\times 7.3}{0.463}$

<u>Roots and Powers (p29-30)</u> ☑

27) Find without using a calculator: a) $\sqrt{121}$ b) $\sqrt[3]{64}$ c) 8^2-2^3 d) Ten thousand as a power of ten.
28) Use a calculator to find: a) 7.5^3 b) $\sqrt{23.04}$ c) $\sqrt[3]{512}$
29) a) What are the three power rules? b) If $f = 7^6 \times 7^4$ and $g = 7^5$, what is $f \div g$?

Algebra — Simplifying Terms

Algebra really terrifies so many people. But honestly, it's not that bad. You just have to make sure you <u>understand and learn</u> these <u>basic rules</u> for dealing with algebraic expressions.

Terms (E)

Before you can do anything else with algebra, you must understand what a term is:

> **A TERM IS A COLLECTION OF NUMBERS, LETTERS AND BRACKETS, ALL MULTIPLIED/DIVIDED TOGETHER**

Terms are separated by <u>+ and − signs</u>. Every term has a + or − attached to the <u>front of it</u>.

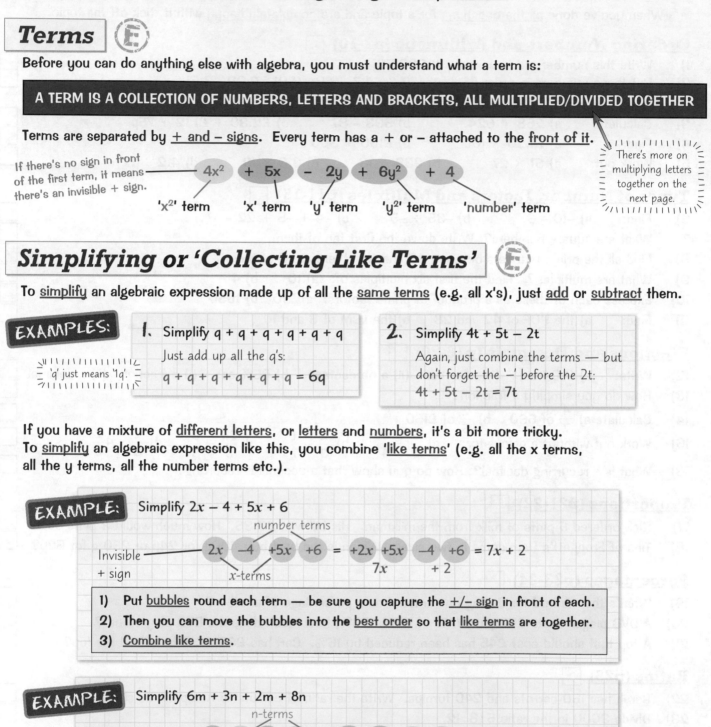

If there's no sign in front of the first term, it means there's an invisible + sign.

$4x^2$ + $5x$ − $2y$ + $6y^2$ + 4

'x^2' term   'x' term   'y' term   'y^2' term   'number' term

There's more on multiplying letters together on the next page.

Simplifying or 'Collecting Like Terms' (E)

To <u>simplify</u> an algebraic expression made up of all the <u>same terms</u> (e.g. all x's), just <u>add</u> or <u>subtract</u> them.

EXAMPLES:

1. Simplify $q + q + q + q + q + q$

Just add up all the q's:

$q + q + q + q + q + q = 6q$

'q' just means '1q'.

2. Simplify $4t + 5t − 2t$

Again, just combine the terms — but don't forget the '−' before the 2t:

$4t + 5t − 2t = 7t$

If you have a mixture of <u>different letters</u>, or <u>letters</u> and <u>numbers</u>, it's a bit more tricky.
To <u>simplify</u> an algebraic expression like this, you combine '<u>like terms</u>' (e.g. all the x terms, all the y terms, all the number terms etc.).

EXAMPLE: Simplify $2x − 4 + 5x + 6$

number terms

Invisible + sign

$2x$ $−4$ $+5x$ $+6$ = $+2x$ $+5x$ $−4$ $+6$ = $7x + 2$

$7x$ $+ 2$

x-terms

1) Put <u>bubbles</u> round each term — be sure you capture the <u>+/− sign</u> in front of each.
2) Then you can move the bubbles into the <u>best order</u> so that <u>like terms</u> are together.
3) <u>Combine like terms</u>.

EXAMPLE: Simplify $6m + 3n + 2m + 8n$

n-terms

Invisible + sign

$6m$ $+3n$ $+2m$ $+8n$ = $+6m$ $+2m$ $+3n$ $+8n$ = $8m + 11n$

$8m$ $+ 11n$

m-terms

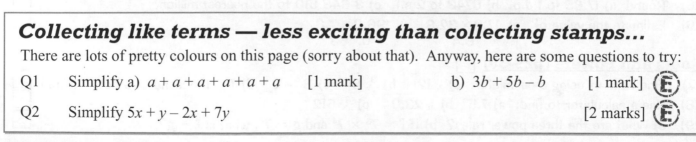

Collecting like terms — less exciting than collecting stamps...

There are lots of pretty colours on this page (sorry about that). Anyway, here are some questions to try:

Q1   Simplify a) $a + a + a + a + a + a$ [1 mark]   b) $3b + 5b − b$ [1 mark] (E)

Q2   Simplify $5x + y − 2x + 7y$ [2 marks] (E)

Algebra — Simplifying Terms

On this page we'll look at some <u>rules</u> that will help you <u>simplify</u> expressions that have <u>letters and numbers multiplied together</u>.

Letters Multiplied Together (D)

Watch out for these combinations of letters in algebra that regularly catch people out:

1) abc means $a \times b \times c$ and 3a means $3 \times a$.   The ×'s are often left out to make it clearer.

2) gn^2 means $g \times n \times n$.   Note that only the n is squared, not the g as well.

3) $(gn)^2$ means $g \times g \times n \times n$.   The brackets mean that <u>BOTH</u> letters are squared.

> There's more on powers on p30.

4) <u>Powers</u> tell you <u>how many</u> letters are multiplied together   — so $r^6 = r \times r \times r \times r \times r \times r$.

5) -3^2 isn't very clear.   It should either be written $(-3)^2 = 9$, or $-(3^2) = -9$ (you'd usually take -3^2 to be -9).

EXAMPLES:

1. Simplify $k \times k \times k \times k$
You have 4 k's multiplied together:
$k \times k \times k \times k = k^4$

Careful — k times itself 4 times is k^4, <u>not</u> 4k (4k means $k + k + k + k$ or $4 \times k$).

2. Simplify $a \times b \times 6$
This one's dead easy — just combine into one term (and put the number at the front):
$a \times b \times 6 = 6ab$

3. Simplify $5s \times 3t$
Multiply the numbers together, then the letters together:
$5s \times 3t = 5 \times 3 \times s \times t = 15st$

Power Rules and Algebra (C)

You can use the <u>power rules</u> from p30 on <u>algebraic expressions</u> too:
1) When <u>multiplying</u>, you <u>add</u> the powers.
2) When <u>dividing</u>, you <u>subtract</u> the powers.

EXAMPLES:

1. Simplify $v^2 \times v^3$
You're multiplying, so <u>add</u> the powers:
$v^2 \times v^3 = v^{2+3} = v^5$

2. Simplify $\dfrac{w^{11}}{w^8}$
This time, you're dividing — so <u>subtract</u> the powers:
$\dfrac{w^{11}}{w^8} = w^{11-8} = w^3$

Ahhh algebra, it's as easy as abc, or (ab)² or something like that...

If you're struggling to use the power rules on algebra, try writing the expressions out in full to see how they work — so $v^2 \times v^3 = v \times v \times v \times v \times v = v^5$. Then have a go at these Exam Practice Questions.

Q1   Simplify a) $e \times e \times e \times e \times e$ [1 mark]   b) $3f \times 6g$ [1 mark] (D)

Q2   Simplify a) $h^4 \times h^5$ [1 mark]   b) $\dfrac{s^9}{s^6}$ [1 mark] (C)

Algebra — Multiplying Out Brackets

If you have an algebraic expression with <u>brackets</u> in, you might be asked to get rid of them by <u>multiplying out the brackets</u>. Your multiplying skills from the previous page will come in handy here.

Multiplying Out Brackets Ⓒ

There are a few <u>key things</u> to remember before you start multiplying out brackets:

1) The thing <u>outside</u> the brackets multiplies <u>each separate term</u> inside the brackets.

2) When <u>letters</u> are multiplied together, they are just written next to each other, e.g. pq.

3) Remember from the previous page that $r \times r = r^2$, and when you multiply terms with numbers <u>and</u> letters in, you multiply the <u>numbers</u> together then the <u>letters</u>.

4) Be very careful with <u>MINUS SIGNS</u> — remember the rules for multiplying them from p11.

EXAMPLES:

1. Expand $3(2x + 5)$

Multiply the $2x$ and 5 inside by the 3 outside:

$3(2x + 5) = (3 \times 2x) + (3 \times 5)$
$= 6x + 15$

2. Expand $4a(3b - 2)$

Multiply the $3b$ and -2 inside by the $4a$ outside:

$4a(3b - 2) = (4a \times 3b) + (4a \times -2)$
$= 12ab - 8a$

$4 \times -2 = -8$

3. Expand $-4(3p^2 - 7q^3)$

Be very careful with the minus signs here:

Note that the minus sign outside the brackets <u>reverses</u> all the signs when you multiply.

$-4(3p^2 - 7q^3) = (-4 \times 3p^2) + (-4 \times -7q^3)$
$= -12p^2 + 28q^3$

4. Expand $2e(e - 3)$

This time, be careful when you multiply $2e$ by e — you'll end up with a $2e^2$:

$2e(e - 3) = (2e \times e) + (2e \times -3)$
$= 2e^2 - 6e$

Collecting Like Terms Ⓒ

If you're given <u>more than one</u> set of brackets to expand like $2(x + 2) + 3(x - 4)$, you'll have to <u>simplify</u> at the end by <u>collecting like terms</u> (see p32).

EXAMPLE:

Expand and simplify $3(x + 2) + 4(4 - x)$.

First, <u>expand</u> each of the brackets separately (as you did above):

$3(x + 2) + 4(4 - x) = (3 \times x) + (3 \times 2) + (4 \times 4) + (4 \times -x)$
$= 3x + 6 + 16 - 4x$

Careful with the negatives here — the 4 and $-x$ multiply to give $-4x$.

Then <u>collect like terms</u> to simplify the expression:

$= 3x - 4x + 6 + 16 = -x + 22$ OR $22 - x$

Go forth and multiply out brackets...

Don't rush when multiplying out brackets — you'll make mistakes and throw away easy marks. Have a go at these Exam Practice Questions to see how it's done (expanding brackets that is, not throwing away marks).

Q1   Expand: a) $-6(3x - 2)$ [1 mark]   b) $3x(x - 5)$ [2 marks] Ⓒ

Q2   Expand and simplify $5(y + 2) + 3(3 - y)$ [2 marks] Ⓒ

Algebra — Taking Out Common Factors

Right, now you know how to expand brackets, it's time to put them back in. This is known as <u>factorising</u>.

Factorising — Putting Brackets In Ⓒ

This is the <u>exact reverse</u> of multiplying out brackets. You have to look for <u>common factors</u> — numbers or letters that go into <u>every term</u>. Here's the method to follow:

1) Take out the <u>biggest number</u> that goes into all the terms.

2) <u>For each letter in turn</u>, take out the <u>highest power</u> (e.g. x, x^2 etc.) that will go into **EVERY** term.

3) Open the brackets and fill in all the bits needed to <u>reproduce each term</u>.

4) <u>Check</u> your answer by <u>multiplying out</u> the brackets again.

> <u>REMEMBER</u>: The bits <u>taken out</u> and put at the front of the brackets are the <u>common factors</u>. The bits <u>inside</u> are what get you back to the <u>original terms</u> when you multiply out again.

Taking Out a Number Ⓒ

If <u>both</u> terms of the expression you're trying to factorise have a <u>number part</u>, you can look for a <u>common factor</u> of both numbers and take it <u>outside</u> the brackets. The common factor is the <u>biggest number</u> that the numbers in <u>both terms</u> divide by.

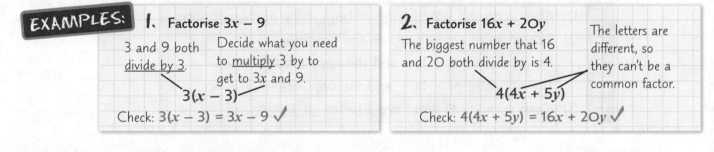

EXAMPLES:

1. Factorise $3x - 9$

3 and 9 both <u>divide by 3</u>.

Decide what you need to <u>multiply</u> 3 by to get to $3x$ and 9.

$$3(x - 3)$$

Check: $3(x - 3) = 3x - 9$ ✓

2. Factorise $16x + 20y$

The biggest number that 16 and 20 both divide by is 4.

The letters are different, so they can't be a common factor.

$$4(4x + 5y)$$

Check: $4(4x + 5y) = 16x + 20y$ ✓

Taking Out a Letter Ⓒ

If the <u>same letter</u> appears in <u>all</u> the terms (but to <u>different powers</u>), you can take out some <u>power</u> of the <u>letter</u> as a <u>common factor</u>. You might be able to take out a <u>number</u> as well.

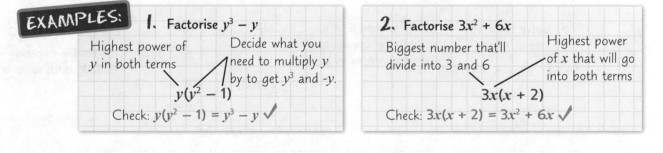

EXAMPLES:

1. Factorise $y^3 - y$

Highest power of y in both terms

Decide what you need to multiply y by to get y^3 and $-y$.

$$y(y^2 - 1)$$

Check: $y(y^2 - 1) = y^3 - y$ ✓

2. Factorise $3x^2 + 6x$

Biggest number that'll divide into 3 and 6

Highest power of x that will go into both terms

$$3x(x + 2)$$

Check: $3x(x + 2) = 3x^2 + 6x$ ✓

Have you got the Common Factor...

Make sure you find <u>all</u> the common factors — if you don't, you won't have <u>fully</u> factorised the expression.

Q1   Factorise: a) $21x - 14y$ [1 mark]   b) $x^2 + 2x$ [1 mark] Ⓒ

Q2   Factorise fully: a) $18a + 12a^2$ [2 marks]   b) $4r^2 - 22rs$ [2 marks] Ⓒ

Solving Equations

'Solving equations' basically means 'find the value of x (or whatever letter is used) that makes the equation true'. To do this, you usually have to rearrange the equation to get x on its own.

The 'Common Sense' Approach (E)

The trick here is to realise that the unknown quantity 'x' is just a number and the 'equation' is a cryptic clue to help you find it.

EXAMPLE: Solve the equation $3x + 4 = 46$. ← This just means 'find the value of x'.

This is what you should say to yourself:

'Something + 4 = 46', hmmm, so that 'something' must be 42.

So that means $3x = 42$, which means '3 × something = 42'.

So it must be 42 ÷ 3 = 14, so $x = 14$.

If you were writing this down in an exam question, just write down the bits in blue.

In other words don't think of it as algebra, but as 'find the mystery number'.

The 'Proper' Way (E)

The 'proper' way to solve equations is to keep rearranging them until you end up with '$x =$' on one side. There are a few important points to remember when rearranging.

Golden Rules

1) Always do the SAME thing to both sides of the equation.
2) To get rid of something, do the opposite.
   The opposite of + is − and the opposite of − is +.
   The opposite of × is ÷ and the opposite of ÷ is ×.
3) Keep going until you have a letter on its own.

EXAMPLES:

1. Solve $x + 7 = 11$.

The opposite of +7 is −7

$x + 7 = 11$

(−7) $x + 7 − 7 = 11 − 7$

$x = 4$

This means 'take away 7 from both sides'.

2. Solve $x − 3 = 7$.

The opposite of −3 is +3

$x − 3 = 7$

(+3) $x − 3 + 3 = 7 + 3$

$x = 10$

3. Solve $5x = 15$.

5x means 5 × x, so do the opposite — divide both sides by 5

$5x = 15$

(÷5) $5x ÷ 5 = 15 ÷ 5$

$x = 3$

4. Solve $\frac{x}{3} = 2$.

$\frac{x}{3}$ means x ÷ 3, so do the opposite — multiply both sides by 3

$\frac{x}{3} = 2$

(×3) $\frac{x}{3} × 3 = 2 × 3$

$x = 6$

Handy hint — x often hides behind the sofa...

It's a good idea to write down what you're doing at every stage — put it in brackets next to the equation (like in the examples above). Try it out on these Exam Practice Questions.

Q1 Solve these equations: a) $x + 2 = 8$ [1 mark] b) $x − 6 = 8$ [1 mark]
  c) $4x = 12$ [1 mark] d) $\frac{x}{5} = 3$ [1 mark] (E)

Solving Equations

You're not done with solving equations yet — not by a long shot. This is where it gets <u>really fun</u>*.

Two-Step Equations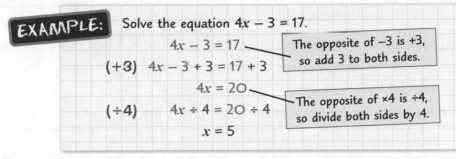

If you come across an equation like $4x + 3 = 19$ (where there's an <u>x-term</u> and a <u>number</u> on the <u>same side</u>), use the methods from the previous page to solve it — just do it in <u>two steps</u>:

1)   <u>Add or subtract</u> the number first.   2)   <u>Multiply or divide</u> to get 'x = '.

EXAMPLE: Solve the equation $4x - 3 = 17$.

$$4x - 3 = 17$$
$$(+3)\quad 4x - 3 + 3 = 17 + 3$$
$$4x = 20$$
$$(\div 4)\quad 4x \div 4 = 20 \div 4$$
$$x = 5$$

The opposite of −3 is +3, so add 3 to both sides.

The opposite of ×4 is ÷4, so divide both sides by 4.

Equations with 'x' on Both Sides D

For equations like $2x + 3 = x + 7$ (where there's an x-term on <u>each side</u>), you have to:

1)   Get all the x's on one side and all the <u>numbers</u> on the other.

2)   <u>Multiply or divide</u> to get 'x = '.

EXAMPLE: Solve the equation $3x + 5 = 5x + 7$.

$$3x + 5 = 5x + 7$$
$$(-3x)\quad 3x + 5 - 3x = 5x + 7 - 3x$$
$$5 = 2x + 7$$
$$(-7)\quad 5 - 7 = 2x + 7 - 7$$
$$-2 = 2x$$
$$(\div 2)\quad -2 \div 2 = 2x \div 2$$
$$-1 = x$$

To get the x's on only one side, subtract 3x from each side.

Now subtract 7 to get the numbers on the other side.

The opposite of ×2 is ÷2, so divide both sides by 2.

Don't be put off by the fact that the x ends up on the right, not the left — $-1 = x$ is exactly the same as $x = -1$.

Equations with Brackets C

If the equation has <u>brackets</u> in, you have to <u>multiply out</u> the brackets (see p34) before solving it as above.

EXAMPLE: Solve the equation $5x + 3 = 4(x + 2)$.

$$5x + 3 = 4(x + 2)$$
$$5x + 3 = 4x + 8$$
$$(-4x)\quad 5x + 3 - 4x = 4x + 8 - 4x$$
$$x + 3 = 8$$
$$(-3)\quad x + 3 - 3 = 8 - 3$$
$$x = 5$$

Multiply out the brackets.

To get the x's on only one side, subtract 4x from each side.

The opposite of +3 is −3, so subtract 3 from each side.

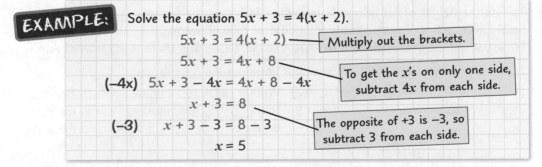

Solving mysteries would be more exciting...

A good thing about solving equations is that you can always check your answer — just put the value of x you've found back into the original equation, and check that it works. Give it a go on these questions.

Q1   Solve $6x - 5 = 3x + 10$ [2 marks] D   Q2   Solve $4(y - 2) = 2y + 6$ [2 marks] C

*Fun not guaranteed. Terms and conditions apply.

Using Formulas

Formulas come up again and again in GCSE Maths, so make sure you're happy with using them.
The first thing you need to be able to do is pretty easy — it's just putting numbers into them.

Putting Numbers into Formulas (E)

You might be given a formula and asked to work out its value when you put in certain numbers.
All you have to do here is follow this method.

> 1) Write out the formula.
>
> 2) Write it again, directly underneath, but substituting numbers for letters on the RHS (right-hand side).
>
> 3) Work it out in stages. Use BODMAS (see p2) to work things out in the right order.
> Write down values for each bit as you go along.
>
> 4) DO NOT attempt to do it all in one go on your calculator — you're more likely to make mistakes.

EXAMPLE: H = 7j − 2k. Find the value of H when j = 4 and k = 5.

H = 7j − 2k ——————— 1) Write out the formula.

H = 7 × 4 − 2 × 5 ——— 2) Write it again, substituting numbers for letters on the RHS.

H = 28 − 10 ——————— 3) Use BODMAS to work things out in the right order —
work out the multiplications first, then do the subtraction.

H = 18

EXAMPLE: The formula for converting from Celsius (C) to Fahrenheit (F) is F = $\frac{9}{5}$C + 32.
Use this formula to convert −10 °C into Fahrenheit.

F = $\frac{9}{5}$C + 32 ——— 1) Write out the formula.

—— 2) Write it again, substituting numbers for letters on the RHS.

F = $\frac{9}{5}$ × −10 + 32 —— 3) Use BODMAS to work things out in the right order —
do the multiplication first, then do the addition.

F = −18 + 32

F = 14   so −10 °C = 14 °F

> Be careful when substituting negative numbers into a formula — just do it step-by-step.

Wordy Formulas (E)

If you're given a formula in words rather than letters, don't panic. You use the exact same method as above.

EXAMPLE: To find the height of his beanstalk in metres, Jack uses the formula:
height = (number of magic beans × amount of rainfall overnight in mm) + 75.
How tall will his beanstalk be if he uses 4 magic beans and there is 30 mm of rain overnight?

height = (number of beans × rainfall) + 75 —— 1) Write out the formula.

height = (4 × 30) + 75 ———————— 2) Write it again, substituting numbers for letters on the RHS.

height = 120 + 75 ———————————— 3) Use BODMAS to work things out in the right order —
do the bit in brackets first, then do the addition.

height = 195 m

Level of grumpiness = number of annoying people ÷ hours of sleep...

If you have more than one number to put into a formula, make sure you put them in the right places in the formula — don't get them mixed up. Have a go at this Exam Practice Question to test your skills.

Q1   Z = 5a + 4b. Find the value of Z when a = 3 and b = 6.   [2 marks] (E)

Making Formulas From Words

Before we get started, there are a few <u>definitions</u> you need to know:

> 1) EXPRESSION — a <u>collection</u> of <u>terms</u> (see p32). Expressions <u>DON'T</u> have an = sign in them.
> 2) EQUATION — an expression with an = sign in it (so you can solve it)
> 3) FORMULA — a <u>rule</u> that helps you work something out (it will also have an = sign in it).

Making a Formula from Given Information (E)

Making <u>formulas</u> from <u>words</u> can be a bit confusing as you're given a lot of <u>information</u> in one go.
You just have to go through it slowly and carefully and <u>extract the maths</u> from it.

EXAMPLE: Tiana is x years old. Leah is 5 years younger than Tiana. Martin is 4 times as old as Tiana.

a) Write an expression for Leah's age in terms of x.

Tiana's age is x

So Leah's age is $x - 5$ ← Leah is 5 years younger, so subtract 5

b) Write an expression for Martin's age in terms of x.

Tiana's age is x

So Martin's age is $4 \times x = 4x$ ← 4 times older

EXAMPLE: Windsurfing lessons cost £15 per hour, plus a fixed fee of £20 for equipment hire. h hours of lessons cost £W. Write a formula for W in terms of h.

$$W = 15h + 20$$

One hour costs 15, so h hours will cost 15 × h

Don't forget to add on the fixed fee (20)

> Because you're asked for a formula, you must include the 'W = ' bit to get full marks (i.e. don't just put 15h + 20).

EXAMPLE: In rugby union, tries score 5 points and conversions score 2 points. In a game, Morgan scores a total of M points, made up of t tries and c conversions. Write a formula for M in terms of t and c.

Tries score 5 points ——— t tries will score 5 × t = 5t points

Conversions score 2 points ——— c conversions will score 2 × c = 2c points

So total points scored are M = 5t + 2c

> Penalties and drop-goals score 3 points (in case you thought I'd forgotten) — but he doesn't score any of these.

Using Your Formula to Solve Equations (D)

Sometimes, you might be asked to <u>use</u> a formula to <u>solve an equation</u>.

EXAMPLE: A decorator uses the formula C = 200r + 150, where C is the cost in £ and r is the number of rooms. Gabrielle spends £950. How many rooms does she have decorated?

$$C = 200r + 150$$ — Write down the formula first.

$$950 = 200r + 150$$ — Replace C with the value given in the question (£950).

(−150) $$950 - 150 = 200r + 150 - 150$$ — Now solve the equation.

$$800 = 200r$$

(÷200) $$800 \div 200 = 200r \div 200$$

$$4 = r$$ So Gabrielle has 4 rooms decorated

In h hours of windsurfing, I fell off 8h times...

You know the drill — learn this page and have a go at this Exam Practice Question (then have a cup of tea).

Q1   The cost of hiring a wallpaper-stripper is £12 per day, plus a deposit of £18.
If the cost for hiring it for d days is £C, find an expression for C in terms of d.   [3 marks] (E)

Rearranging Formulas

The <u>subject</u> of a formula is the letter <u>on its own</u> before the = (so x is the subject of x = 2y + 3z).

Changing the Subject of a Formula ©

<u>Rearranging formulas</u> means making a different letter the <u>subject</u>, e.g. getting 'y = ' from 'x = 3y + 2'
— you have to get the subject <u>on its own</u>. Fortunately, you can use the <u>same methods</u> that you used for
<u>solving equations</u> (see p36-37) — here's a quick reminder:

Golden Rules
1) Always do the <u>SAME thing</u> to <u>both sides of the formula</u>.
2) To get rid of something, do the <u>opposite</u>.
   The opposite of + is – and the opposite of – is +.
   The opposite of × is ÷ and the opposite of ÷ is ×.
3) Keep going until you have the letter you want <u>on its own</u>.

EXAMPLE: Rearrange p = q + 12 to make q the subject of the formula.

$$p = q + 12$$

The opposite of +12 is –12, so take away 12 from both sides.

$$(-12) \quad p - 12 = q + 12 - 12$$
$$p - 12 = q \quad OR \quad q = p - 12$$

EXAMPLE: Rearrange a = 3b + 4 to make b the subject of the formula.

$$a = 3b + 4$$

The opposite of +4 is –4, so take away 4 from both sides.

$$(-4) \quad a - 4 = 3b + 4 - 4$$
$$a - 4 = 3b$$

The opposite of ×3 is ÷3, so divide both sides by 3.

$$(\div 3) \quad (a - 4) \div 3 = 3b \div 3$$
$$\frac{a-4}{3} = b \quad OR \quad b = \frac{a-4}{3}$$

Careful here — you divide the <u>whole side</u> by 3, not just one term.

EXAMPLE: Rearrange m = $\frac{n}{4}$ – 7 to make n the subject of the formula.

$$m = \frac{n}{4} - 7$$

The opposite of –7 is +7, so add 7 to both sides.

$$(+7) \quad m + 7 = \frac{n}{4} - 7 + 7$$
$$m + 7 = \frac{n}{4}$$

The opposite of ÷4 is ×4, so multiply both sides by 4.

$$(\times 4) \quad (m + 7) \times 4 = \frac{n}{4} \times 4$$
$$4(m + 7) = n \quad OR \quad n = 4m + 28$$

If I could rearrange my subjects I'd have Maths all day every day...

This page is really just like solving equations — so if you learn the method for one, you know
the method for the other. What a bonus. It's like buy-one-get-one-free but more mathsy.

Q1 Rearrange $u = \frac{v}{3} - 2$ to make v the subject of the formula. [2 marks] ©

Q2 Rearrange $c = 6d - 12$ to make d the subject of the formula. [2 marks] ©

Number Patterns and Sequences

Sequences are just patterns of numbers or shapes that follow a rule. You need to be able to spot what the rule is.

Finding Number Patterns

The trick to finding the rule for number patterns is to write down what you have to do to get from one number to the next in the gaps between the numbers. There are 2 main types to look out for:

1) Add or subtract the same number

E.g.   2   5   8   11   14   ...   30   24   18   12   ...
            +3   +3   +3   +3   +3   –6   –6   –6   –6

The RULE: 'Add 3 to the previous term'   'Subtract 6 from the previous term'

2) Multiply or divide by the same number each time

E.g.   2   6   18   54   ...   40 000   4000   400   40   ...
           ×3   ×3   ×3   ×3   ÷10   ÷10   ÷10   ÷10

The RULE: 'Multiply the previous term by 3'   'Divide the previous term by 10'

You might sometimes get patterns that follow a different rule — for example, you might have to add or subtract a changing number each time, or add together the two previous terms. You probably don't need to worry about this, but if it comes up, just describe the pattern and use your rule to find the next term.

Shape Patterns

If you have a pattern of shapes, you need to be able to continue the pattern. You might also have to find the rule for the pattern to work out how many shapes there'll be in a later pattern.

EXAMPLE: On the right, there are some patterns made of circles.
a) Draw the next pattern in the sequence.
b) Work out how many circles there will be in the 10th pattern.

a) Just continue the pattern —
each 'leg' increases by one circle.

In an exam question, you might be given a table and asked to complete it.

b) Set up a table to find the rule:

Pattern number	1	2	3	4	5	6	7	8	9	10
Number of circles	1	3	5	7	9	11	13	15	17	19

The rule is 'add 2 to the previous term'.

So just keep on adding 2 to extend the table until you get to the 10th term — which is **19**.

Knitting patterns follow the rule knit one, purl one...

Remember, you always need to work out what to do to get from one term to the next — that's the rule.

Q1   A sequence starts 27, 22, 17, 12. Write down the next term in the sequence
and explain how you worked it out.
[2 marks]

Number Patterns and Sequences

If the examiners are feeling mean, they might ask you to "find an <u>expression</u> for the <u>nth term</u> of a sequence" — this is a rule with n in, like 5n – 3. It gives <u>every term in a sequence</u> when you put in different values for n.

Finding the nth Term of a Sequence Ⓒ

This method works for sequences with a <u>common difference</u> — where you <u>add</u> or <u>subtract</u> the <u>same number</u> each time (i.e. the difference between each pair of terms is the <u>same</u>).

EXAMPLE:

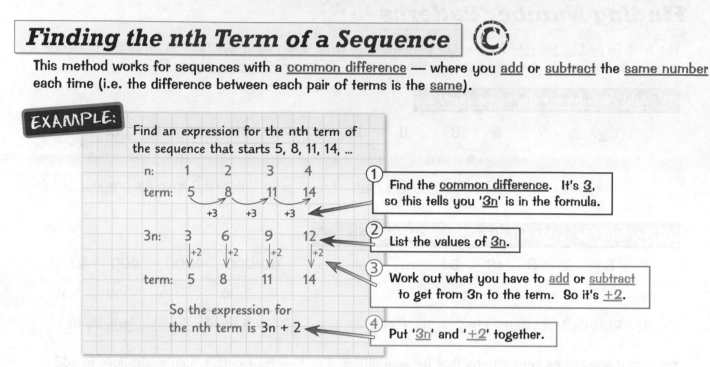

Find an expression for the nth term of the sequence that starts 5, 8, 11, 14, …

| n: | 1 | 2 | 3 | 4 |

term: 5 → 8 → 11 → 14
  +3   +3   +3

① Find the <u>common difference</u>. It's <u>3</u>, so this tells you '<u>3n</u>' is in the formula.

3n: 3 6 9 12
  +2 +2 +2 +2
term: 5 8 11 14

② List the values of <u>3n</u>.

③ Work out what you have to <u>add</u> or <u>subtract</u> to get from 3n to the term. So it's <u>+2</u>.

So the expression for the nth term is 3n + 2

④ Put '<u>3n</u>' and '<u>+2</u>' together.

<u>Check</u> your formula by putting the first few values of n back in:
n = 1 gives 3n + 2 = 3 + 2 = 5 ✓
n = 2 gives 3n + 2 = 6 + 2 = 8 ✓

Deciding if a Term is in a Sequence Ⓒ

You might be given the nth term and asked if a <u>certain value</u> is in the sequence. The trick here is to <u>set the expression equal to that value</u> and solve to find n. If n is a <u>whole number</u>, the value is <u>in</u> the sequence.

EXAMPLE: A sequence is given by the rule 6n – 2.

a) Find the 6th term in the sequence.

Just put n = 6 into the expression:
(6 × 6) – 2 = 36 – 2
           = 34

b) Is 45 a term in this sequence?

Set it equal to 45…   6n – 2 = 45
                       6n = 47   …and solve for n.
                       n = 47 ÷ 6 = 7.8333…

n is not a whole number, so 45 is <u>not</u> in the sequence 6n – 2.

It might be even <u>easier</u> to decide if a number is in a sequence or not — for example, if the sequence was all <u>odd numbers</u>, there's <u>no way</u> that an <u>even number</u> could be in the sequence. You just have to use your common sense — e.g. if all the terms in the sequence ended in <u>3</u> or <u>8</u>, 44 would <u>not</u> be in the sequence.

If I've told you n times, I've told you n + 1 times — learn this page…

This is a bit trickier than the other stuff on sequences, but at least there's a way to check your answer. The only way to get good at finding the rule is practice — so have a go at this Exam Practice Question.

Q1 A sequence starts 2, 9, 16, 23, …

  a) Find an expression for the nth term of the sequence. [2 marks]

  b) Use your expression to find the 8th term in the sequence. [1 mark] Ⓒ

  c) Is 63 a term in the sequence? Explain your answer. [2 marks]

Trial and Improvement

Trial and improvement is a way of finding an approximate solution to an equation that's too hard to be solved using normal methods. You'll always be told WHEN to use trial and improvement — don't go using it willy-nilly.

Keep Trying Different Values in the Equation Ⓒ

The basic idea of trial and improvement is to keep trying different values of x that are getting closer and closer to the solution. Here's the method to follow:

> **STEP 1:** Put 2 values into the equation that give opposite cases (one too big, one too small).
>
> **STEP 2:** Choose the next value between the two opposite cases.
>
> **STEP 3:** Repeat STEP 2 until you have two numbers:
> - both to 1 d.p.
> - differing by 1 in the last digit (e.g. 4.3 and 4.4).
>
> These are the two possible answers.
>
> **STEP 4:** Take the exact middle value to decide which one it is.

If you had to find a solution to 2 d.p., the method's just the same — except you'll end up with 2 numbers to 2 d.p. instead of 1 d.p.

Put Your Working in a Table Ⓒ

It's a good idea to keep track of your working in a table — see the example below.

EXAMPLE:

The solution to the equation $x^3 + 9x = 40$ lies between 2 and 3.
Use trial and improvement to find the solution to this equation to 1 d.p.

	x	$x^3 + 9x$		
STEP 1: Put in 2 and 3 first (given in question)	2	26	Too small	... so solution is between 2 and 3
	3	54	Too big	
STEP 2: Try 2.5	2.5	38.125	Too small	... so solution is between 2.5 and 3
STEP 3: Try 2.7	2.7	43.983	Too big	... so solution is between 2.5 and 2.7
Try 2.6	2.6	40.976	Too big	... so solution is between 2.5 and 2.6
STEP 4: Take the exact middle value	2.55	39.531375	Too small	So the answer to 1 d.p. has to be either 2.5 or 2.6...

... so solution is between 2.55 and 2.6, so $x = 2.6$ to 1 d.p.

Make sure you show all your working — otherwise the examiner won't be able to tell what method you've used and you'll lose marks.

Trial and improvement — not a good strategy for lion-taming...

Not the most exciting page in the world sorry — but it is a good way of picking up easy marks in the exam just by putting some numbers into equations. Try it out on these Exam Practice Questions:

Q1 $x^3 + 6x = 69$ has a solution between 3 and 4.
Use trial and improvement to find this solution to 1 d.p.   [4 marks] Ⓒ

Q2 $x^3 - 12x = 100$ has a solution between 5 and 6.
Use trial and improvement to find this solution to 1 d.p.   [4 marks] Ⓒ

Inequalities

Inequalities are a bit tricky, but once you've learned the tricks involved, most of the algebra for them is identical to ordinary equations (have a look back at pages 36-37 if you need a reminder).

The Inequality Symbols (C)

> I > All of you.

>	means 'Greater than'	≥	means 'Greater than or equal to'
<	means 'Less than'	≤	means 'Less than or equal to'

REMEMBER — the one at the BIG end is BIGGEST so x > 4 and 4 < x both say: 'x is greater than 4'.

EXAMPLE: x is an integer such that $-4 < x \leq 3$. Write down all possible values of x.

Work out what each bit of the inequality is telling you:

$-4 < x$ means 'x is greater than -4',

and $x \leq 3$ means 'x is less than or equal to 3'.

Now just write down all the values that x can take:

-3, -2, -1, 0, 1, 2, 3

Remember, integers are just whole numbers (+ve and −ve, including O).

−4 isn't included because of the < but 3 is included because of the ≤.

You Can Show Inequalities on Number Lines (C)

Drawing inequalities on a number line is dead easy — all you have to remember is that you use an open circle (O) for > or < and a coloured-in circle (●) for ≥ or ≤.

EXAMPLE: Show the inequality $-4 < x \leq 3$ on a number line.

Closed circle because 3 is included.

Open circle because −4 isn't included.

```
←—————+——O——+——+——+——+——O——+——+——●——+——+——→
      -5  -4  -3  -2  -1   O   1   2   3   4   5
```

Algebra with Inequalities (C)

Algebra with inequalities isn't actually that bad because inequalities are just like regular equations — you can use all the normal rules of algebra (there is one exception, but you don't need to worry about it).

EXAMPLES:

1. Solve $3x - 2 \leq 13$.

Just solve it like an equation — but leave the inequality sign in your answer:

$(+2)$ $\quad 3x - 2 + 2 \leq 13 + 2$

$\qquad\qquad 3x \leq 15$

$(\div 3)$ $\quad 3x \div 3 \leq 15 \div 3$

$\qquad\qquad x \leq 5$

2. Solve $2x + 7 > x + 11$.

Again, solve it like an equation:

(-7) $\quad 2x + 7 - 7 > x + 11 - 7$

$\qquad\qquad 2x > x + 4$

$(-x)$ $\quad 2x - x > x + 4 - x$

$\qquad\qquad x > 4$

Oh the injustice of inequalities...

The exception that I mentioned is that if you multiply or divide by a negative number, you have to flip the inequality sign round (so < becomes > and ≤ becomes ≥). But you shouldn't need to do this in the exam.

Q1 $\quad n$ is an integer such that $-1 \leq n < 5$. Write down all the possible values of n. $\qquad$ [2 marks] (C)

Q2 $\quad$ Solve the following inequalities: a) $4x + 3 < 27$ $\quad$ [2 marks] $\quad$ b) $4x \geq 18 - 2x$ $\quad$ [2 marks] (C)

Revision Questions for Section Two

There was a lot of <u>nasty algebra</u> in that section — let's see how much you remember.
* Try these questions and <u>tick off each one</u> when you <u>get it right</u>.
* When you've done <u>all the questions</u> for a topic and are <u>completely happy</u> with it, tick off the topic.

Algebra (p32-35) ☑

1) Simplify: a) $e + e + e$ b) $4f + 5f - f$
2) Simplify: a) $2x + 3y + 5x - 4y$ b) $11a + 2 - 8a + 7$
3) Simplify: a) $m \times m \times m$ b) $p \times q \times 7$ c) $2x \times 9y$
4) Simplify: a) $g^5 \times g^6$ b) $c^{15} \div c^{12}$
5) Expand: a) $6(x + 3)$ b) $-3(3x - 4)$ c) $x(5 - x)$
6) Expand and simplify $4(3 + 5x) - 2(7x + 6)$
7) What is factorising?
8) Factorise: a) $8x + 24$ b) $18x + 27y$ c) $5x^2 + 15x$

Solving Equations (p36-37) ☑

9) Solve: a) $x + 9 = 16$ b) $x - 4 = 12$ c) $6x = 18$
10) Solve a) $4x + 3 = 19$ b) $3x + 6 = x + 10$ c) $3(x + 2) = 5x$

Formulas (p38-40) ☑

11) $Q = 5r + 6s$. Work out the value of Q when $r = -2$ and $s = 3$.
12) Imran buys d DVDs and c CDs. DVDs cost £7 each and CDs cost £5 each. He spends £P in total. Write a formula for P in terms of d and c.
13) Hiring ice skates costs £3 per hour plus a deposit of £5. Lily paid £11. How long did she hire the skates for?
14) Rearrange the formula $W = 4v + 5$ to make v the subject.

Number Patterns and Sequences (p41-42) ☑

15) For each of the following sequences, find the next term and write down the rule you used.
a) 3, 10, 17, 24, ... b) 1, 4, 16, 64, ... c) 2, 5, 7, 12, ...
16) Find an expression for the nth term of the sequence that starts 4, 10, 16, 22, ...
17) Is 34 a term in the sequence given by the expression $7n - 1$?

Trial and Improvement (p43) ☑

18) What is trial and improvement?
19) Given that $x^3 + 8x = 103$ has a solution between 4 and 5, use trial and improvement to find this solution to 1 d.p.
20) Given that $x^3 - 3x = 41$ has a solution between 3 and 4, use trial and improvement to find this solution to 1 d.p.

Inequalities (p44) ☑

21) Write the following inequalities out in words: a) $x > -7$ b) $x \leq 6$
22) $0 < k \leq 7$. Find all the possible integer values of k.
23) Solve the following inequalities: a) $x + 4 < 14$ b) $x - 11 > 3$ c) $7x \geq 21$
24) Solve the inequality $3x + 5 \leq 26$.

Coordinates and Midpoints

What could be more fun than points in one quadrant? Points in <u>four quadrants</u>, that's what...

The Four Quadrants (F)

A graph has <u>four different quadrants</u> (regions).

The top-right region is the easiest because
<u>ALL THE COORDINATES IN IT ARE POSITIVE</u>.

You have to be careful in the <u>other regions</u> though,
because the x- and y- coordinates could be <u>negative</u>,
and that makes life much more difficult.

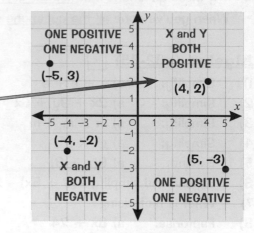

THREE IMPORTANT POINTS ABOUT COORDINATES:

1) The coordinates are always in <u>ALPHABETICAL ORDER</u>, x then y.   **(x , y)**

2) x is always the flat axis going <u>ACROSS</u> the page.
   In other words <u>'x is a..cross'</u>   Get it — x is a '×'.   (Hilarious isn't it)

3) Remember it's always <u>IN THE HOUSE</u> (→) and then <u>UP THE STAIRS</u> (↑)
   so it's <u>ALONG first</u> and <u>then UP</u>, i.e. x-coordinate first, and then y-coordinate.

The Midpoint of a Line (C)

The '<u>MIDPOINT OF A LINE SEGMENT</u>' is the <u>POINT THAT'S BANG IN THE MIDDLE</u> of it.

Finding the coordinates of a midpoint is pretty easy.
<u>LEARN THESE THREE STEPS</u>...

1) Find the <u>average</u> of the <u>x-coordinates</u>.
2) Find the <u>average</u> of the <u>y-coordinates</u>.
3) Plonk them in <u>brackets</u>.

Midpoint
of Jeff

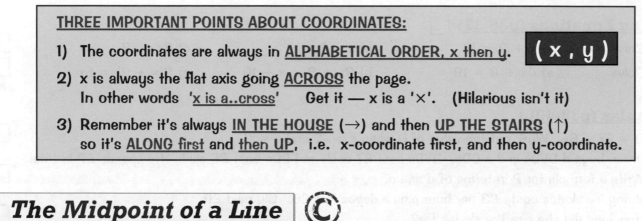

EXAMPLE: P and Q have coordinates (1, 2) and (6, 6).
Find the <u>midpoint</u> of the line PQ.

Average of x-coordinates $= \dfrac{1+6}{2} = 3.5$

Average of y-coordinates $= \dfrac{2+6}{2} = 4$

Coordinates of midpoint = (3.5, 4)

But what if you live in a bungalow...

Learn the 3 points for getting *x* and *y* the right way round and then try these questions.

Q1   a) Plot point A(−3, 2) and point B(3, 5) on a grid.   [2 marks] (F)

       b) Find the coordinates of the midpoint of AB.   [2 marks] (C)

Straight-Line Graphs

If you thought I-spy was a fun game, wait 'til you play 'recognise the straight-line graph from the equation'.

Horizontal and Vertical lines: 'x = a' and 'y = a' (E)

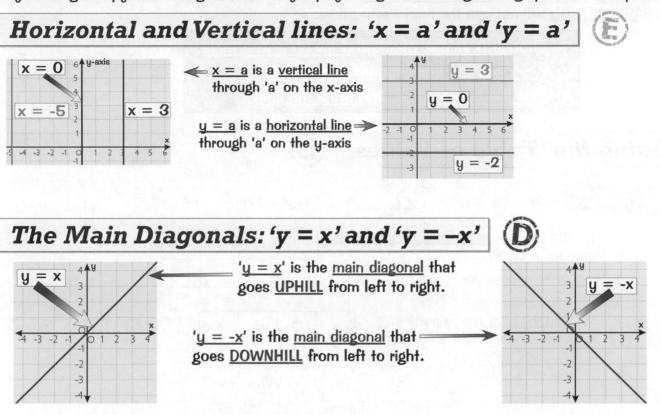

x = a is a <u>vertical line</u> through 'a' on the x-axis

y = a is a <u>horizontal line</u> through 'a' on the y-axis

The Main Diagonals: 'y = x' and 'y = –x' (D)

'y = x' is the <u>main diagonal</u> that goes <u>UPHILL</u> from left to right.

'y = -x' is the <u>main diagonal</u> that goes <u>DOWNHILL</u> from left to right.

Other Lines Through the Origin: 'y = ax' and 'y = –ax' (D)

<u>y = ax</u> and <u>y = -ax</u> are the equations for **A SLOPING LINE THROUGH THE ORIGIN**.

The value of '<u>a</u>' (known as the <u>gradient</u>) tells you the steepness of the line. The bigger 'a' is, the steeper the slope. A <u>MINUS SIGN</u> tells you it slopes <u>DOWNHILL</u>.

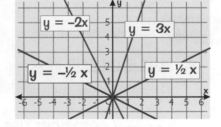

Learn to Spot Straight Lines from their Equations (D)

All straight-line equations just contain '<u>something x</u>, <u>something y</u> and <u>a number</u>'.

<u>Straight lines:</u>		<u>NOT straight lines:</u>	
$x - y = 0$	$y = 2 + 3x$	$y = x^3 + 3$	$\frac{1}{y} + \frac{1}{x} = 2$
$2y - 4x = 7$	$4x - 3 = 5y$	$x^2 = 4 - y$	$xy + 3 = 0$

There's more on x^2 graphs on page 54.

My favourite line's y = 3x — it gets the ladies every time...

It's definitely worth learning all the graphs above. Once you've done that, test yourself with this question.

Q1   On a grid with x-axis from –5 to 5 and y-axis from –5 to 5, draw these lines:

   a) $y = -1$ b) $y = -x$ c) $x = 2$ [3 marks] (D)

Plotting Straight-Line Graphs

You're likely to be asked to <u>DRAW THE GRAPH</u> of an equation in the exam.
This <u>EASY METHOD</u> will net you the marks every time:

> 1) Choose <u>3 values of x</u> and <u>draw up a wee table</u>,
> 2) <u>Work out the corresponding y-values</u>,
> 3) <u>Plot the coordinates</u>, and <u>draw the line</u>.

You might get lucky and be <u>given</u> a table in an exam question. Don't worry if it contains <u>5 or 6 values</u>.

Doing the 'Table of Values' (D)

EXAMPLE: Draw the graph of <u>$y = 2x - 3$</u> for values of x from −2 to 4.

1. <u>Choose 3 easy x-values for your table:</u>
   Use x-values from the grid you're given.
   Avoid negative ones if you can.

x	0	2	4
y			

2. <u>Find the y-values</u> by putting each x-value into the equation:

x	0	2	4
y	−3	1	5

When $x = 0$,
$y = 2x - 3$
$= (2 \times 0) - 3 = -3$

When $x = 4$,
$y = 2x - 3$
$= (2 \times 4) - 3 = 5$

Plotting the Points and Drawing the Graph (D)

EXAMPLE: ...continued from above.

3. <u>PLOT EACH PAIR</u> of x- and y- values from your table.

   The table gives the coordinates (0, −3), (2, 1) and (4, 5).

   Now draw a <u>STRAIGHT LINE</u> through your points.

   > If one point looks a bit wacky, check 2 things:
   > – the <u>y-value</u> you worked out in the table
   > – that you've <u>plotted</u> it properly.

(4, 5) (2, 1) Dead straight line (0, −3)

Careful plotting — the key to straight lines and world domination...

If the examiners are feeling mean, they'll give you an equation like $3x + y = 5$ to plot, making finding the y-values a tad trickier. Just substitute the x-value and find the y-value that makes the equation true.
E.g. when $x = 1$, $3x + y = 5 \rightarrow (3 \times 1) + y = 5 \rightarrow 3 + y = 5 \rightarrow y = 2$.

Q1 Draw the graph of $y = x + 4$ for values of x from −6 to 2. [3 marks] (D)

Q2 Draw the graph of $y + 3x = 2$ for values of x from −2 to 2. [3 marks] (D)

Straight-Line Graphs — Gradients

Time to hit the slopes. Well, find them anyway...

Finding the Gradient Ⓒ

The <u>gradient</u> of a line is a measure of its <u>slope</u>. The <u>bigger</u> the number, the <u>steeper</u> the line.

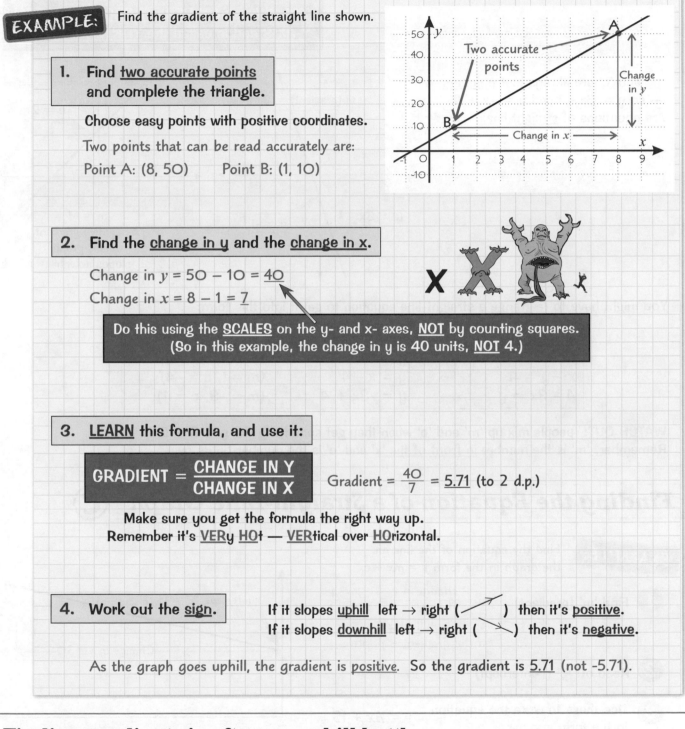

EXAMPLE: Find the gradient of the straight line shown.

1. Find <u>two accurate points</u> and complete the triangle.

   Choose easy points with positive coordinates.

   Two points that can be read accurately are:

   Point A: (8, 50)   Point B: (1, 10)

2. Find the <u>change in y</u> and the <u>change in x</u>.

   Change in y = 50 − 10 = <u>40</u>
   Change in x = 8 − 1 = <u>7</u>

   > Do this using the <u>SCALES</u> on the y- and x- axes, <u>NOT</u> by counting squares.
   > (So in this example, the change in y is 40 units, <u>NOT</u> 4.)

3. <u>LEARN</u> this formula, and use it:

   $$\text{GRADIENT} = \frac{\text{CHANGE IN Y}}{\text{CHANGE IN X}}$$

   Gradient = $\frac{40}{7}$ = <u>5.71</u> (to 2 d.p.)

   Make sure you get the formula the right way up.
   Remember it's <u>VER</u>y <u>HO</u>t — <u>VER</u>tical over <u>HO</u>rizontal.

4. Work out the <u>sign</u>.

   If it slopes <u>uphill</u> left → right (⟋) then it's <u>positive</u>.
   If it slopes <u>downhill</u> left → right (⟍) then it's <u>negative</u>.

   As the graph goes uphill, the gradient is <u>positive</u>. So the gradient is <u>5.71</u> (not -5.71).

Finding gradients is often an uphill battle...

Learn the four steps for finding a gradient then have a bash at this Exam Practice Question. Take care — you might not be able to pick two points with nice, positive coordinates. Fun times ahoy.

Q1   Find the gradient of the line shown.   [2 marks] Ⓒ

Straight-Line Graphs — "y = mx + c"

This sounds a bit scary, but give it a go and you might like it.

y = mx + c is the Equation of a Straight Line (C)

$y = mx + c$ is the general equation for a straight-line graph, and you need to remember:

> 'm' is equal to the <u>GRADIENT</u> of the graph
>
> 'c' is the value <u>WHERE IT CROSSES THE Y-AXIS</u> and is called the <u>Y-INTERCEPT</u>.

'<u>m</u>' and '<u>c</u>' are always just <u>numbers</u> — so $y = 3x - 1$ and $y = -x + 2$ are equations of <u>straight lines</u>. ⟶

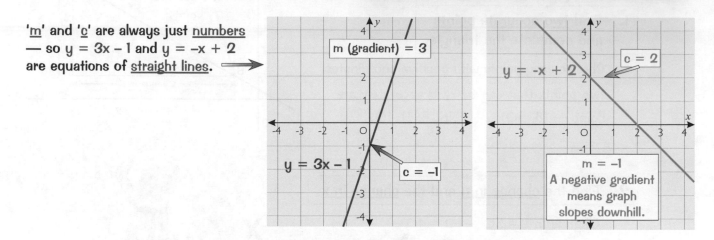

m (gradient) = 3

$y = 3x - 1$

c = −1

$y = -x + 2$

c = 2

m = −1

A negative gradient means graph slopes downhill.

You might have to <u>rearrange</u> a straight-line equation to get it into this form:

<u>Straight line:</u>		<u>Rearranged into 'y = mx + c'</u>	
$y = 2 + 3x$	→	$y = 3x + 2$	(m = 3, c = 2)
$x - y = 4$	→	$y = x - 4$	(m = 1, c = −4)
$4 - 3x = y$	→	$y = -3x + 4$	(m = −3, c = 4)

<u>WATCH OUT</u>: people mix up 'm' and 'c' when they get something like $y = 5 + 2x$. Remember, 'm' is the number <u>in front of the 'x'</u> and 'c' is the number <u>on its own</u>.

Finding the Equation of a Straight-Line Graph (C)

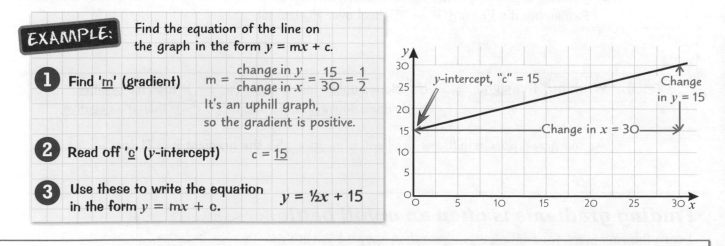

EXAMPLE: Find the equation of the line on the graph in the form $y = mx + c$.

① Find '<u>m</u>' (gradient) $m = \dfrac{\text{change in } y}{\text{change in } x} = \dfrac{15}{30} = \dfrac{1}{2}$

It's an uphill graph, so the gradient is positive.

② Read off '<u>c</u>' (y-intercept) $c = \underline{15}$

③ Use these to write the equation in the form $y = mx + c$. $y = \frac{1}{2}x + 15$

y-intercept, "c" = 15

Change in y = 15

Change in x = 30

Remember y = mx + c — it'll keep you on the straight and narrow...

Remember what 'm' and 'c' mean, and make sure you've identified them correctly.

Q1   What is the gradient of the line with equation $y = 4 - 2x$?   [1 mark] (C)

Travel Graphs

Ah, what could be better than some nice mid-section travel graphs? OK, so a picture of Jennifer Lawrence might be better. Or Hugh Jackman. But the section isn't called 'Hollywood Hotties' is it...

Distance-Time Graphs Ⓓ

1) The graph **GOING UP** means it's travelling **AWAY**. The graph **COMING DOWN** means it's **COMING BACK AGAIN**.

2) At any point, **GRADIENT = SPEED**, but watch out for the UNITS.

3) The **STEEPER** the graph, the **FASTER** it's going

4) **FLAT SECTIONS** are where it is **STOPPED**.

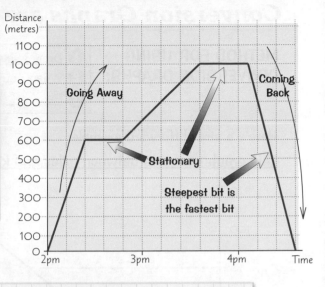

EXAMPLE: Henry went out for a ride on his bike. After a while he got a puncture and stopped to fix it. This graph shows the first part of Henry's journey.

a) **What time did Henry leave home?**

He left home at the point where the line starts.

At 8:15

b) **How far did Henry cycle before getting a puncture?**

The horizontal part of the graph is where Henry stopped.

12 km

c) **What was Henry's speed before getting a puncture?**

The gradient of this part of the graph gives you Henry's speed before he got a puncture.

$$\text{speed} = \text{gradient} = \frac{\text{change in } y}{\text{change in } x}$$

$$= \frac{12 \text{ km}}{0.5 \text{ hours}}$$

$$= 24 \text{ km/h}$$

d) At 9:30 Henry cycled straight home. It took him 45 minutes. **Complete the graph to show this.**

Henry gets home 45 minutes after 9:30, which is 10:15. So complete the graph with a straight line.

D-T Graphs — filled with highs and lows, just like life...

The only way to get good at distance-time graphs is to practise, practise, practise...

Q1 a) Using the graph above, how long did Henry stop for? [1 mark] Ⓓ

    b) How far from home was Henry at 8:30? [1 mark] Ⓓ

Conversion Graphs

In the exam you're likely to get a graph which converts something like <u>£ to dollars</u> or <u>mph to km/h</u>.

Conversion Graphs are Easy to Use Ⓔ

METHOD FOR USING CONVERSION GRAPHS:

❶ <u>Draw a line</u> from a value on <u>one axis</u>.
❷ When you hit the LINE, <u>change direction</u> and go straight to <u>the other axis</u>.
❸ <u>Read off the value</u> from this axis. The two values are <u>equivalent</u>.

Here's a straightforward example:

<u>This graph converts between miles and kilometres</u>

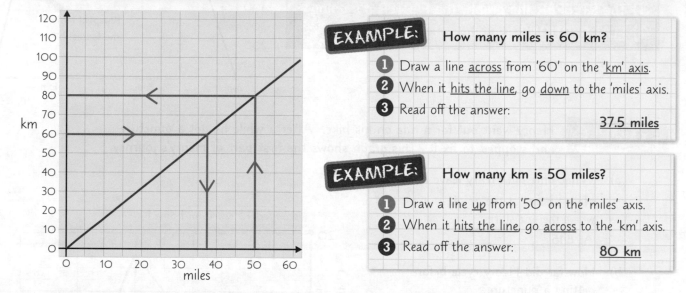

EXAMPLE: How many miles is 60 km?

❶ Draw a line <u>across</u> from '60' on the '<u>km</u>' axis.
❷ When it <u>hits the line</u>, go <u>down</u> to the 'miles' axis.
❸ Read off the answer:
                                    <u>37.5 miles</u>

EXAMPLE: How many km is 50 miles?

❶ Draw a line <u>up</u> from '50' on the 'miles' axis.
❷ When it <u>hits the line</u>, go <u>across</u> to the 'km' axis.
❸ Read off the answer:
                                    <u>80 km</u>

Using Conversion Graphs to Answer Harder Questions

Conversion graphs are so <u>simple</u> to use that the examiners often wrap them up in tricky questions.

EXAMPLE: Sam went on holiday to Florida and paid $360 for a camera. The same camera in Manchester costs £250. Where was the camera cheaper? Show your working. Ⓓ

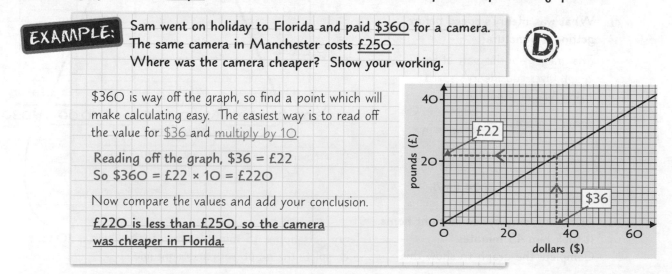

$360 is way off the graph, so find a point which will make calculating easy. The easiest way is to read off the value for $36 and multiply by 10.

Reading off the graph, $36 = £22
So $360 = £22 × 10 = £220

Now compare the values and add your conclusion.

<u>£220 is less than £250, so the camera was cheaper in Florida.</u>

Learn how to convert graph questions into marks...

Draw your conversion lines on the graph in the exam. If all else fails, this might get you a mark.

Q1   The distance between Tokyo and London is 6000 miles.
      Use the graph at the top of the page to estimate this distance in km.
                                                                    [2 marks] Ⓓ

Real-Life Graphs

Graphs can be drawn for just about anything. I can't show you all the possibilities due to health and safety concerns about back injuries resulting from the weight of the book. But here are some useful examples.

Graphs Can Show How Much You'll Pay (D)

Graphs are great for showing how much you'll be charged for using a service or buying multiple items.

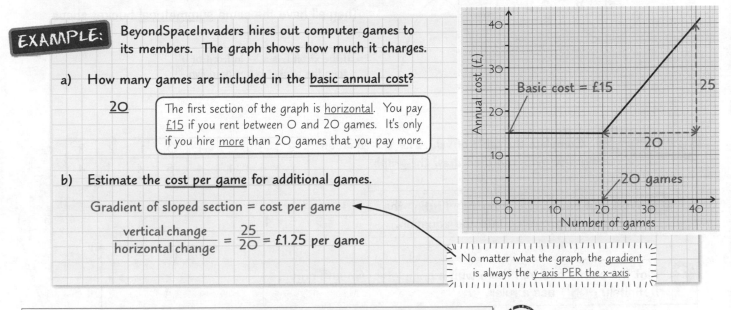

EXAMPLE: BeyondSpaceInvaders hires out computer games to its members. The graph shows how much it charges.

a) How many games are included in the basic annual cost?

20

> The first section of the graph is horizontal. You pay £15 if you rent between 0 and 20 games. It's only if you hire more than 20 games that you pay more.

b) Estimate the cost per game for additional games.

Gradient of sloped section = cost per game

$\dfrac{\text{vertical change}}{\text{horizontal change}} = \dfrac{25}{20} = £1.25$ per game

No matter what the graph, the gradient is always the y-axis PER the x-axis.

Graphs Can Show Other Changes Too (D)

Graphs aren't always about money. They can also show things like water depth or temperature.

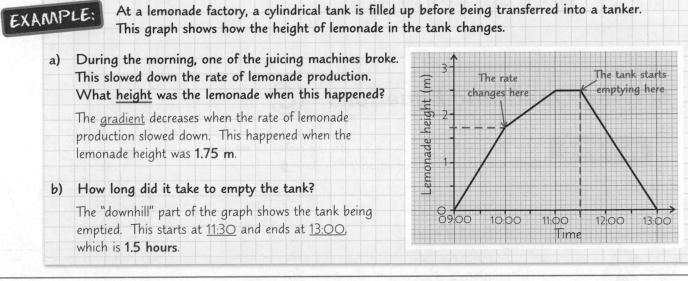

EXAMPLE: At a lemonade factory, a cylindrical tank is filled up before being transferred into a tanker. This graph shows how the height of lemonade in the tank changes.

a) During the morning, one of the juicing machines broke. This slowed down the rate of lemonade production. What height was the lemonade when this happened?

The gradient decreases when the rate of lemonade production slowed down. This happened when the lemonade height was **1.75 m**.

b) How long did it take to empty the tank?

The "downhill" part of the graph shows the tank being emptied. This starts at 11:30 and ends at 13:00, which is **1.5 hours**.

Rate of filling brain with Maths > Rate of forgetting (hopefully)...

If you're stumped on a graph question, have a think about what the gradient means — the thing on the vertical axis PER the thing on the horizontal axis. This might help you.

Q1 Sore Thumbs Games Club has no membership fee but charges £1.50 to hire a game.
    a) Draw a graph to show the cost of hiring up to 20 games from this club. [3 marks] (D)
    b) Robert plans to hire 5 games over the year. Which club will charge the least, Sore Thumbs or BeyondSpaceInvaders (above)? Explain your answer. [3 marks]

Q2 Using the graph at the bottom of the page, estimate the times when there was 2.2 m of lemonade in the tank. [2 marks] (D)

Quadratic Graphs

Enough of straight lines. You now get to graduate to lovely, smooth curves. Quadratic ones to be precise.

Drawing a Quadratic Graph Ⓒ

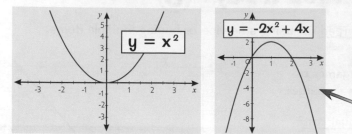

Quadratic graphs are of the form $\underline{y = \text{anything with } x^2}$ (but not higher powers of x).

They all have the same <u>symmetrical</u> bucket shape.

If the x^2 bit has a '–' in front of it then the bucket is <u>upside down</u>.

EXAMPLE: Complete the table of values for the equation $y = x^2 - 5$ and then draw the graph.

x	-3	-2	-1	0	1	2	3
y	4	-1	-4	-5	-4	-1	4

① Work out each <u>y-value</u> by <u>substituting</u> the corresponding <u>x-value</u> into the equation.

$y = (-3)^2 - 5$
$= 9 - 5 = 4$

$y = (2)^2 - 5$
$= 4 - 5 = -1$

② Plot the points and join them with a <u>completely smooth curve</u>. Definitely <u>DON'T</u> use a ruler.

<u>NEVER EVER</u> let one point drag your line off in some ridiculous direction. When a graph is generated from an equation, you never get spikes or lumps — only <u>MISTAKES</u>.

This point is <u>obviously wrong</u>

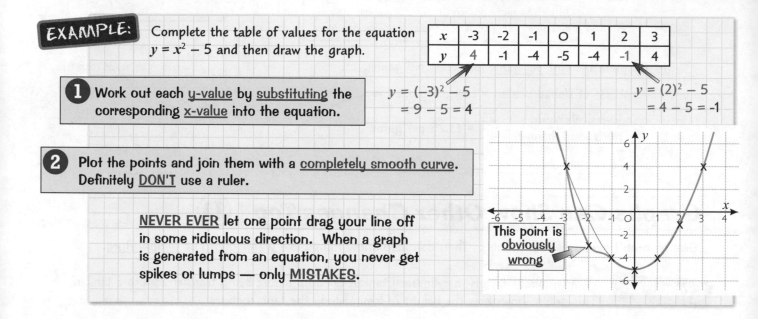

Solving Quadratic Equations Ⓒ

EXAMPLE: Use the graph of $\underline{y = 2x^2 - 3x}$ (on the right) to estimate both solutions to the equation $2x^2 - 3x = \underline{5}$.

$2x^2 - 3x = 5$ is what you get when you put $\underline{y = 5}$ into the graph's equation, so:

1) <u>Draw</u> a line at $\underline{y = 5}$.

2) Read the <u>x-values</u> where the curve <u>crosses</u> this line.

The solutions are about $\underline{x = -1}$ and $\underline{x = 2.5}$.

Quadratic equations usually have <u>2 solutions</u>.

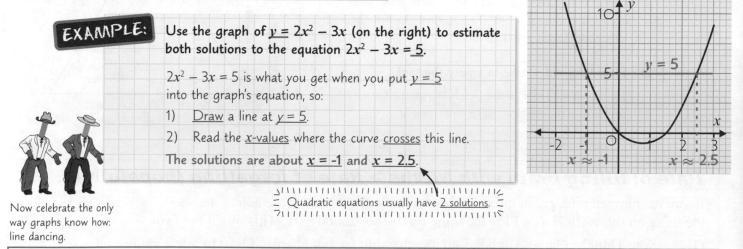

Now celebrate the only way graphs know how: line dancing.

How refreshing — a page on graphs. Not seen one of those in a while...

You know the deal by now — learn what's on this page, then treat yourself to answering the question below.

Q1 a) Draw the graph of $y = x^2 - 1$ for values of x between –3 and 3. [4 marks] Ⓒ

 b) Use your graph to estimate the solutions to $5 = x^2 - 1$. [2 marks] Ⓒ

Revision Questions for Section Three

Well, that wraps up <u>Section Three</u> — time to put yourself to the test and find out <u>how much you really know</u>.

- Try these questions and <u>tick off each one</u> when you <u>get it right</u>.
- When you've done <u>all the questions</u> for a topic and are <u>completely happy</u> with it, tick off the topic.

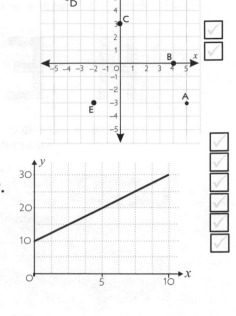

Coordinates and Midpoints (p46) ☑

1) Give the coordinates of points A to E in the diagram on the right.
2) Find the midpoint of a line segment with endpoints B and C.

Straight-Line Graphs and their Gradients (p47-50) ☑

3) Sketch the lines a) $y = -x$, b) $y = -4$, c) $x = 2$
4) What does a straight-line equation look like?
5) Use the table of three values method to draw the graph $y = 2x + 3$.
6) What does a line with a negative gradient look like?
7) Find the gradient of the line on the right.
8) What do 'm' and 'c' represent in $y = mx + c$?

Travel Graphs (p51) ☑

9) What does a horizontal line mean on a distance-time graph?
10) The graph on the right shows Ben's car journey to the supermarket and home again.
   a) Did he drive faster on his way to the supermarket or on his way home?
   b) How long did he spend at the supermarket?

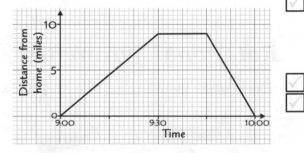

Conversion and Real-Life Graphs (p52-53) ☑

11) This graph shows the monthly cost of a mobile phone contract.
   a) What is the basic monthly fee?
   b) How many minutes does the monthly fee include?
   c) Mary uses her phone for 35 minutes one month. What will her bill be?
   d) Stuart is charged £13.50 one month. How long did he use his phone for?
   e) Estimate the cost per minute for additional minutes. Give your answer to the nearest 1 p.

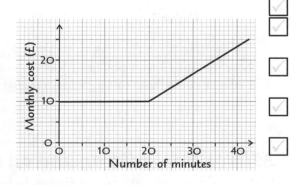

Quadratic Graphs (p54) ☑

12) Describe the shapes of the graphs $y = x^2 - 8$ and $y = -x^2 + 2$.
13) Plot the graph $y = x^2 + 2x$ for values of x between −3 and 3, and use it to solve $2 = x^2 + 2x$.

Symmetry

There are two types of <u>symmetry</u> that you need to know about — <u>line symmetry</u> and <u>rotational symmetry</u>.

Line Symmetry (F)

This is where you draw one or more <u>MIRROR LINES</u> across a shape and both sides will <u>fold exactly</u> together.

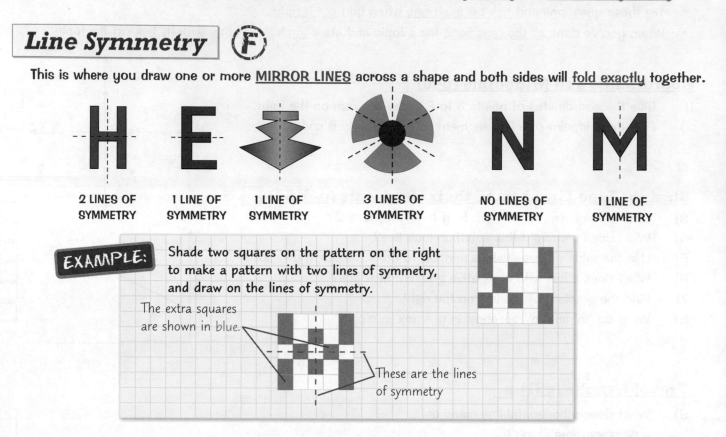

| 2 LINES OF SYMMETRY | 1 LINE OF SYMMETRY | 1 LINE OF SYMMETRY | 3 LINES OF SYMMETRY | NO LINES OF SYMMETRY | 1 LINE OF SYMMETRY |

EXAMPLE: Shade two squares on the pattern on the right to make a pattern with two lines of symmetry, and draw on the lines of symmetry.

The extra squares are shown in blue.

These are the lines of symmetry

Rotational Symmetry (F)

This is where you can <u>rotate</u> the shape into different positions that <u>look exactly the same</u>.

| Order 1 | Order 2 | Order 2 | Order 3 | Order 4 |

The <u>ORDER OF ROTATIONAL SYMMETRY</u> is the posh way of saying: 'how many different positions look the same'. You should say the Z-shape above has '<u>rotational symmetry order 2</u>'.

> When a shape has <u>only 1 position</u> you can <u>either</u> say that it has 'rotational symmetry <u>order 1</u>' or that it has '<u>NO</u> rotational symmetry'.

Mirror line, mirror line on the wall...

Make sure you learn the two different types of symmetry, and dazzle your friends by spotting them in everyday shapes like road signs, warning signs and letters. Once you're happy with symmetry, try this Exam Practice Question.

Q1   Make two copies of the pattern above right.
      a)   Shade two squares to make a pattern with one line of symmetry.   [2 marks]
      b)   Shade two squares to make a pattern with rotational symmetry of order 2.   [2 marks] (F)

Symmetry and Tessellations

One more handy bit of advice about <u>symmetry</u>, then it's on to <u>tessellations</u>. Can you bear the suspense?

Tracing Paper (F)

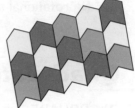

<u>Tracing paper</u> makes symmetry questions a lot easier.

1) For <u>REFLECTIONS</u>, trace one side of the drawing and the mirror line too.
   Then <u>turn the paper over and line up the mirror line</u> in its original position.

2) For <u>ROTATIONS</u>, just swizzle the paper round. It's really good for <u>finding the centre of rotation</u> (see p80) as well as the <u>order of rotational symmetry</u>.

3) You can use tracing paper in the <u>EXAM</u>.

Tessellations (F)

<u>Tessellations are **tiling patterns with no gaps**</u>

Some shapes <u>don't</u> tessellate — there'll be <u>gaps</u> in the pattern like this:

These are regular octagons — there's more about them on p77.

In the <u>exam</u>, you might have to <u>show</u> how a shape tessellates — this just means that you have to use the shape to <u>draw a pattern</u> with <u>no gaps</u> in it. Sometimes you might have to <u>rotate</u> a shape to make them fit together.

EXAMPLE: Show how the shape on the right tessellates. You must draw at least 6 shapes.

Just fit the shapes together, making sure you don't leave any gaps between them:

2, 4, 6, 8 — show how you can tessellate...

Drawing patterns is a nice way to pick up marks — just don't get carried away and make up your own shapes (the examiners don't like it). Try this Exam Practice Question to show off your tessellating skills.

Q1   On squared paper, show how the shape on the right tessellates. You must draw at least 6 shapes.   [2 marks] (F)

Properties of 2D Shapes

This page is jam-packed with details about <u>triangles</u> and <u>quadrilaterals</u> — and you need to learn them all.

Triangles (Three-Sided Shapes)

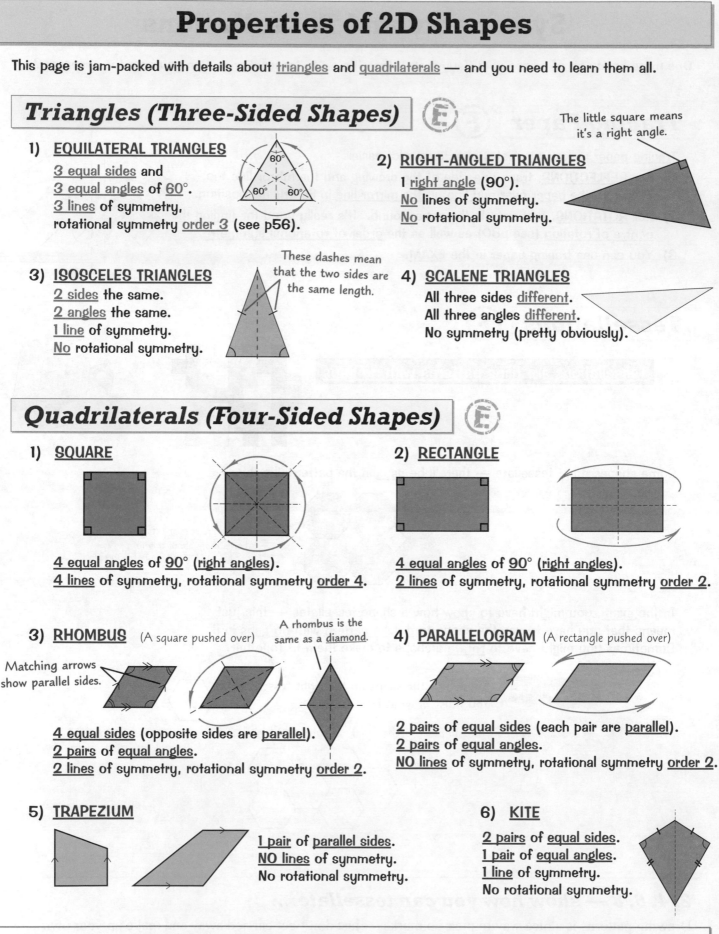

1) **EQUILATERAL TRIANGLES**

   <u>3 equal sides</u> and
   <u>3 equal angles</u> of <u>60°</u>.
   <u>3 lines</u> of symmetry,
   rotational symmetry <u>order 3</u> (see p56).

2) **RIGHT-ANGLED TRIANGLES**

   The little square means it's a right angle.

   1 <u>right angle</u> (90°).
   <u>No</u> lines of symmetry.
   <u>No</u> rotational symmetry.

3) **ISOSCELES TRIANGLES**

   <u>2 sides</u> the same.
   <u>2 angles</u> the same.
   <u>1 line</u> of symmetry.
   <u>No</u> rotational symmetry.

   These dashes mean that the two sides are the same length.

4) **SCALENE TRIANGLES**

   All three sides <u>different</u>.
   All three angles <u>different</u>.
   No symmetry (pretty obviously).

Quadrilaterals (Four-Sided Shapes)

1) **SQUARE**

   <u>4 equal angles</u> of <u>90°</u> (<u>right angles</u>).
   <u>4 lines</u> of symmetry, rotational symmetry <u>order 4</u>.

2) **RECTANGLE**

   <u>4 equal angles</u> of <u>90°</u> (<u>right angles</u>).
   <u>2 lines</u> of symmetry, rotational symmetry <u>order 2</u>.

3) **RHOMBUS**   (A square pushed over)

   A rhombus is the same as a <u>diamond</u>.

   Matching arrows show parallel sides.

   <u>4 equal sides</u> (opposite sides are <u>parallel</u>).
   <u>2 pairs</u> of <u>equal angles</u>.
   <u>2 lines</u> of symmetry, rotational symmetry <u>order 2</u>.

4) **PARALLELOGRAM**   (A rectangle pushed over)

   <u>2 pairs</u> of <u>equal sides</u> (each pair are <u>parallel</u>).
   <u>2 pairs</u> of <u>equal angles</u>.
   <u>NO lines</u> of symmetry, rotational symmetry <u>order 2</u>.

5) **TRAPEZIUM**

   <u>1 pair</u> of <u>parallel sides</u>.
   <u>NO lines</u> of symmetry.
   No rotational symmetry.

6) **KITE**

   <u>2 pairs</u> of <u>equal sides</u>.
   <u>1 pair</u> of <u>equal angles</u>.
   <u>1 line</u> of symmetry.
   No rotational symmetry.

Kite facts — 2 pairs of equal sides, 1 line of symmetry, Gemini...

Learn the <u>names</u> (and how to spell them) and <u>properties</u> of all the shapes on this page, then try this question:

Q1 A quadrilateral has all 4 sides the same length and two pairs of equal angles.
    Identify the quadrilateral, and write down its order of rotational symmetry.   [2 marks]

Congruence and Similarity

Shapes can be <u>similar</u> or <u>congruent</u>. And I bet you really want to know what that means —
I can already picture your eager face. Well, lucky you — I've written a page all about it.

Congruent — Same Shape, Same Size (F)

<u>Congruence</u> is another ridiculous maths word which sounds really complicated when it's not:

> If two shapes are **CONGRUENT**, they are **EXACTLY THE SAME**
> — the **SAME SIZE** and the **SAME SHAPE**.

There ain't room for the two of us in this town, pal.

These shapes are all <u>congruent</u>:

Note — you can have <u>mirror images</u> or <u>rotations</u>.

EXAMPLE: Two of the triangles below are congruent.
Write down the letters of the congruent triangles.

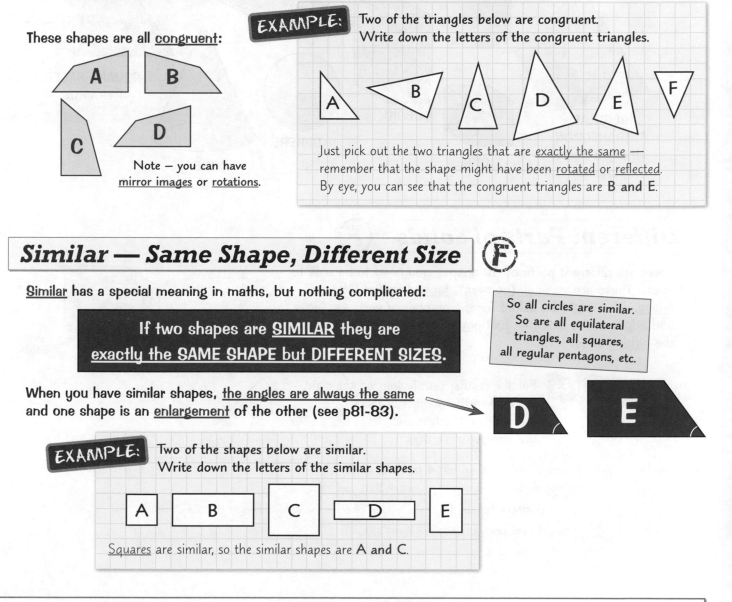

Just pick out the two triangles that are <u>exactly the same</u> —
remember that the shape might have been <u>rotated</u> or <u>reflected</u>.
By eye, you can see that the congruent triangles are **B and E**.

Similar — Same Shape, Different Size (F)

<u>Similar</u> has a special meaning in maths, but nothing complicated:

> If two shapes are **SIMILAR** they are
> exactly the **SAME SHAPE** but **DIFFERENT SIZES**.

So all circles are similar.
So are all equilateral
triangles, all squares,
all regular pentagons, etc.

When you have similar shapes, <u>the angles are always the same</u>
and one shape is an <u>enlargement</u> of the other (see p81-83).

EXAMPLE: Two of the shapes below are similar.
Write down the letters of the similar shapes.

<u>Squares</u> are similar, so the similar shapes are **A and C**.

Butter and margarine — similar products...

To help remember the difference between similarity and congruence,
think '<u>similar siblings, congruent clones</u>' — siblings are alike but not
exactly the same, clones are identical.

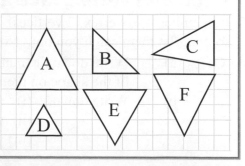

Q1 From the set of triangles on the right, write down the letters of:
a) two congruent triangles [1 mark]
b) two similar triangles [1 mark] (F)

3D Shapes

I was going to make some pop-out <u>3D shapes</u> to put on this page, but I couldn't find the scissors and sticky tape. Sorry. Still, you need to learn it all though — so chin up and learn the page.

Eight Solids to Learn Ⓖ

<u>3D shapes</u> are <u>solid shapes</u>. These are the ones you need to know:

There's more about prisms on p67.

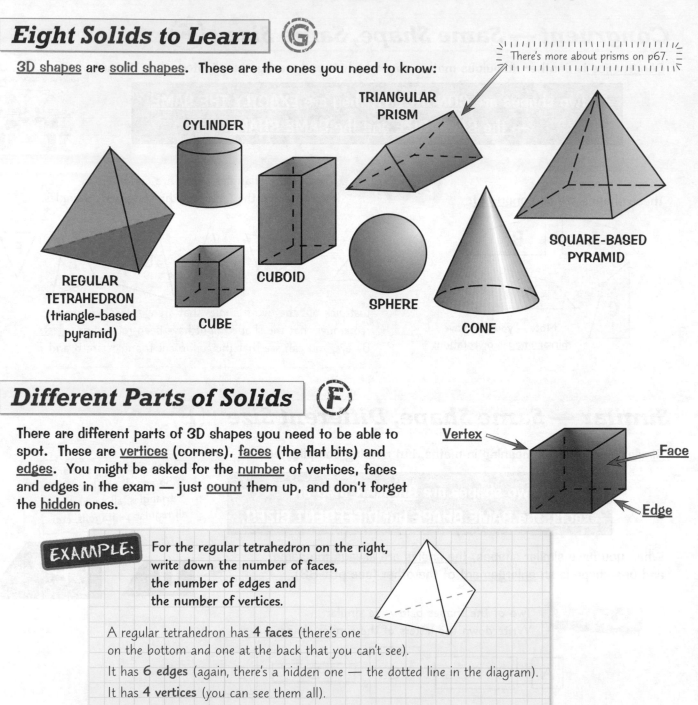

CYLINDER

TRIANGULAR PRISM

SQUARE-BASED PYRAMID

REGULAR TETRAHEDRON (triangle-based pyramid)

CUBE

CUBOID

SPHERE

CONE

Different Parts of Solids Ⓕ

There are different parts of 3D shapes you need to be able to spot. These are <u>vertices</u> (corners), <u>faces</u> (the flat bits) and <u>edges</u>. You might be asked for the <u>number</u> of vertices, faces and edges in the exam — just <u>count</u> them up, and don't forget the <u>hidden</u> ones.

<u>Vertex</u> <u>Face</u> <u>Edge</u>

EXAMPLE: For the regular tetrahedron on the right, write down the number of faces, the number of edges and the number of vertices.

A regular tetrahedron has **4 faces** (there's one on the bottom and one at the back that you can't see).

It has **6 edges** (again, there's a hidden one — the dotted line in the diagram).

It has **4 vertices** (you can see them all).

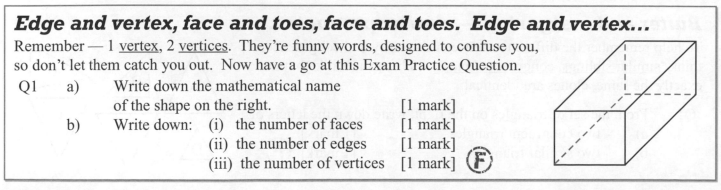

Edge and vertex, face and toes, face and toes. Edge and vertex...

Remember — 1 <u>vertex</u>, 2 <u>vertices</u>. They're funny words, designed to confuse you, so don't let them catch you out. Now have a go at this Exam Practice Question.

Q1   a)   Write down the mathematical name of the shape on the right. [1 mark]

      b)   Write down: (i) the number of faces [1 mark]

                    (ii) the number of edges [1 mark]

                    (iii) the number of vertices [1 mark] Ⓕ

Projections

Projections are just different views of a 3D solid shape — looking at it from the front, the side and the top.

The Three Different Projections Ⓓ

There are three different types of projections — front elevations, side elevations and plans (elevation is just another word for projection).

❶ FRONT ELEVATION
— the view you'd see from directly in front (in the direction of the arrow)

❷ PLAN
— the view you'd see from directly above

❸ SIDE ELEVATION
— the view you'd see from directly to one side

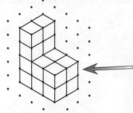

Don't be thrown if you're given a diagram on isometric (dotty) paper like this — it works in just the same way.
If you have to draw shapes on isometric paper, just join the dots.
You should only draw vertical and diagonal lines (no horizontal lines).

Drawing Projections Ⓓ

EXAMPLES:

1. The front elevation and plan view of a shape are shown below. Sketch the solid shape.

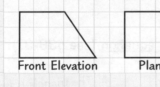

Front Elevation   Plan View

Just piece together the original shape from the information given — here you get a prism in the shape of the front elevation.

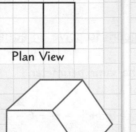

2. a) On the cm square grid, draw the side elevation of the prism from the direction of the arrow.

b) Draw a plan of the prism on the grid.

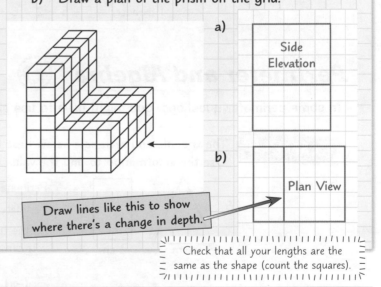

a) Side Elevation

b) Plan View

Draw lines like this to show where there's a change in depth.

Check that all your lengths are the same as the shape (count the squares).

Projections — enough to send you dotty...

Projection questions aren't too bad — just take your time and sketch the diagrams carefully. Watch out for questions on isometric paper — they might look confusing, but they can actually be easier than other questions.

Q1   For the shape on the right, draw:
    a) The front elevation (from the direction of the arrow), [2 marks]
    b) The side elevation, [2 marks]
    c) The plan view. [2 marks] Ⓓ

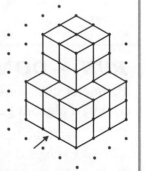

Perimeters

Perimeter is the <u>distance</u> all the way around the <u>outside</u> of a <u>2D shape</u>. It's pretty straightforward if you use the <u>big blob method</u>. So pay attention — this could be easy marks.

Perimeter — Distance Around the Edge of a Shape (F)

To find a <u>perimeter</u>, you <u>add up</u> the <u>lengths</u> of all the sides,
but the only <u>reliable</u> way to make sure you get <u>all</u> the sides is this:

> 1) Put a <u>BIG BLOB</u> at one corner and then go around the shape.
>
> 2) Write down the <u>LENGTH</u> of every side as you go along.
>
> 3) Even sides that seem to have <u>NO LENGTH GIVEN</u>
>   — you must <u>work them out</u>.
>
> 4) Keep going until you get back to the <u>BIG BLOB</u>.

You must choose yourself a blob
and it must also choose you.
It will then be yours for life.

Yes, I know you think it's <u>yet another fussy method</u>, but believe me, it's so easy to miss a side otherwise.

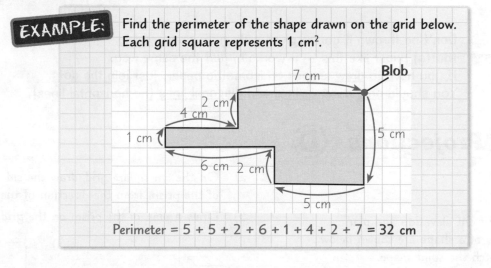

EXAMPLE: Find the perimeter of the shape drawn on the grid below.
Each grid square represents 1 cm².

Blob

7 cm

2 cm

4 cm

1 cm

5 cm

6 cm 2 cm

5 cm

Perimeter = 5 + 5 + 2 + 6 + 1 + 4 + 2 + 7 = **32 cm**

Perimeter and Algebra (D)

In some perimeter questions, you might have to use <u>algebra</u> to <u>solve an equation</u> (see p36-37).

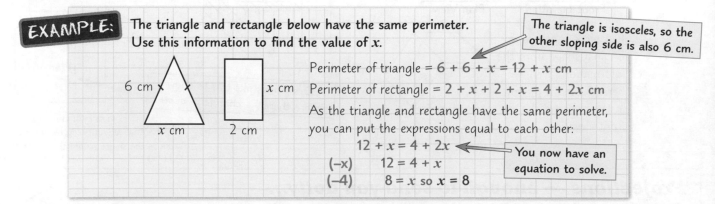

EXAMPLE: The triangle and rectangle below have the same perimeter.
Use this information to find the value of x.

The triangle is isosceles, so the other sloping side is also 6 cm.

6 cm

x cm

x cm   2 cm

Perimeter of triangle = $6 + 6 + x = 12 + x$ cm
Perimeter of rectangle = $2 + x + 2 + x = 4 + 2x$ cm

As the triangle and rectangle have the same perimeter,
you can put the expressions equal to each other:

$$12 + x = 4 + 2x$$

You now have an equation to solve.

$(-x)$ $12 = 4 + x$
(-4) $8 = x$ so $x = 8$

RUN — DON'T WALK from... the BIG BLOB...

Perimeter questions can either be dead easy (if it's on a cm grid) or pretty darn tricky (if algebra gets involved) — make sure you can do both types of question.

Q1   Find the perimeter of the shaded shape.
      Each grid square represents 1 cm².   [1 mark] (F)

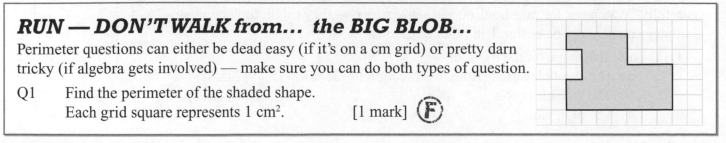

Areas

On this page are four area formulas you need to learn — rectangles, triangles, parallelograms and trapeziums. Remember that area is measured in square units (e.g. cm², m² or km²).

You Must Learn These Four Area Formulas (D)

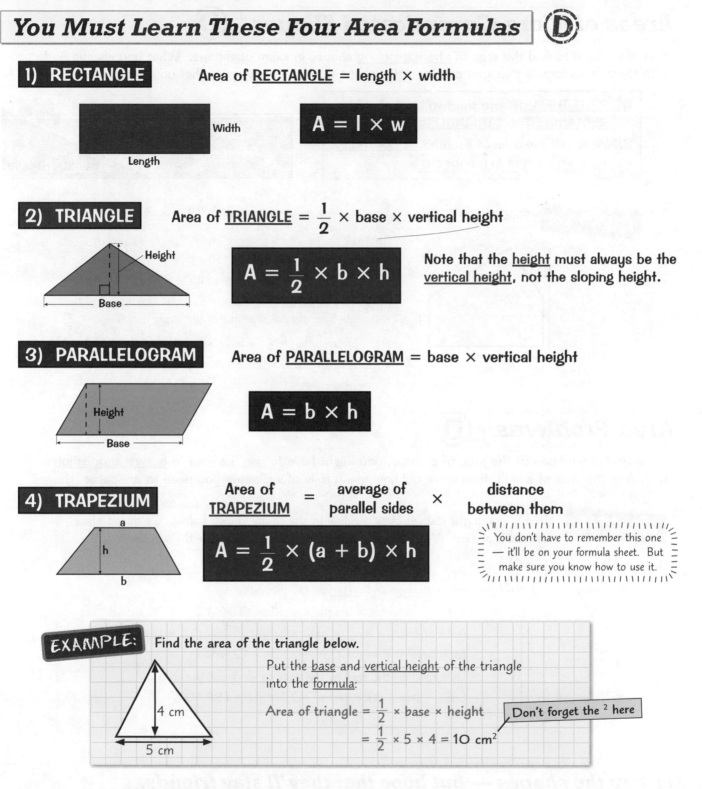

1) RECTANGLE

Area of RECTANGLE = length × width

Width

Length

$$A = l \times w$$

2) TRIANGLE

Area of TRIANGLE = $\frac{1}{2}$ × base × vertical height

Height

Base

$$A = \frac{1}{2} \times b \times h$$

Note that the height must always be the vertical height, not the sloping height.

3) PARALLELOGRAM

Area of PARALLELOGRAM = base × vertical height

Height

Base

$$A = b \times h$$

4) TRAPEZIUM

Area of TRAPEZIUM = average of parallel sides × distance between them

a

h

b

$$A = \frac{1}{2} \times (a + b) \times h$$

You don't have to remember this one — it'll be on your formula sheet. But make sure you know how to use it.

EXAMPLE: Find the area of the triangle below.

Put the base and vertical height of the triangle into the formula:

4 cm

5 cm

Area of triangle = $\frac{1}{2}$ × base × height

= $\frac{1}{2}$ × 5 × 4 = 10 cm²

Don't forget the ² here

No jokes about my vertical height please...

Not much to say about this page really — LEARN the formulas and then try these Exam Practice Questions.

Q1   A rectangular sandpit measures 3 m by 1.8 m. Find its area.   [2 marks] (D)

Q2   Find the area of the trapezium on the right.   [2 marks] (D)

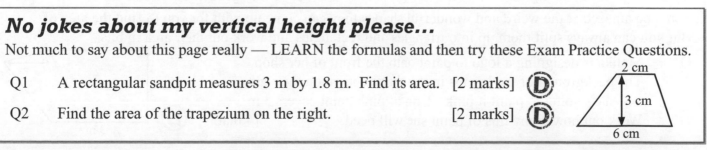

2 cm

3 cm

6 cm

Areas

Make sure you know the <u>formulas</u> for finding the area of <u>rectangles</u> and <u>triangles</u> —
you're going to need them again here.

Areas of More Complicated Shapes Ⓓ

You often have to find the area of <u>strange-looking</u> shapes in exam questions. What you always find
with these questions is that you can break the shape up into <u>simpler ones</u> that you can deal with.

> 1) <u>SPLIT THEM UP</u> into the two basic shapes:
>   <u>RECTANGLES</u> and <u>TRIANGLES</u>.
> 2) Work out the area of each bit <u>SEPARATELY</u>.
> 3) Then <u>ADD THEM ALL TOGETHER</u>.

Basic Rectangle

Basic Triangle

EXAMPLE: Find the area of the shape below.

Split the shape into a <u>triangle</u> and <u>rectangle</u> as shown
and work out the <u>area</u> of each shape:

Area of rectangle = length × width = 8 × 3 = 24 cm²

To find the <u>height</u> of the triangle, subtract the height of the
rectangle from the total height of the shape (so 6 − 3 = 3).

Area of triangle $= \frac{1}{2} \times$ base × height $= \frac{1}{2} \times 8 \times 3 = 12$ cm²

Total area of shape = 24 + 12 = 36 cm²

6 cm, 3 cm, 8 cm

Area Problems Ⓓ

Once you've worked out the <u>area</u> of a shape, you might have to <u>use</u> the area to <u>answer a question</u>
(e.g. find the area of a wall, then work out how many rolls of wallpaper you need to wallpaper it).

EXAMPLE: Greg is making a stained-glass window in the shape shown below. Coloured glass
costs £82 per m². Work out the cost of the glass needed for the window.

First, work out the area of the shape by splitting it
into a <u>triangle</u> and <u>rectangle</u> (as shown):

Area of rectangle = length × width = 0.8 × 1.2 = 0.96 m²

Area of triangle $= \frac{1}{2} \times$ base × height $= \frac{1}{2} \times 0.8 \times 0.6 = 0.24$ m²

Total area of shape = 0.96 + 0.24 = 1.2 m²

Then <u>multiply</u> the <u>area</u> by the <u>price</u> to work out the cost:

Cost = area × price per m² = 1.2 × 82 = £98.40

0.6 m, 1.2 m, 0.8 m

Split up the shapes — but hope that they'll stay friends...

You'd be amazed at the weird and wonderful shapes they can come up with for you to find the area of.
But you can always split them up into triangles and rectangles and work out their area in bits.

Q1   Malika is designing a logo to paint onto the front of her shop.
The background of the logo is in the shape shown on the right,
and she's going to paint it pink. 1 tin of pink paint covers 3 m².
Work out how many tins of paint she will need.   [4 marks] Ⓓ

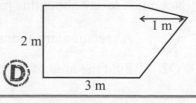

2 m, 1 m, 3 m

Circles

There's a surprising number of <u>circle terms</u> you need to know — don't mix them up. Oh, and it's probably best to have a snack before starting this page. All the talk of <u>pi</u> can make you a bit peckish.

Radius and Diameter (F)

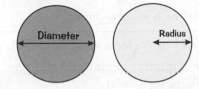

The <u>DIAMETER</u> goes <u>right across</u> the circle, passing through the <u>centre</u>.
The <u>RADIUS</u> goes from the <u>centre</u> of the circle to any point on the <u>edge</u>.

> The <u>DIAMETER IS EXACTLY DOUBLE THE RADIUS</u>

So if the radius is 4 cm, the diameter is 8 cm,
and if the diameter is 24 m, the radius is 12 m.

Area, Circumference and π (D)

There are two more important formulas for you to <u>learn</u> here — <u>circumference</u> and <u>area</u> of a circle.
Circumference is the distance round the outside of the circle (its <u>perimeter</u>).

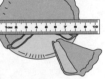

1) <u>CIRCUMFERENCE</u> $= \pi \times$ diameter
   $= \pi \times$ radius $\times 2$
   (as diameter = 2 × radius)

$$C = \pi \times D \text{ or } C = 2 \times \pi \times r$$

2) <u>AREA</u> $= \pi \times$ (radius)^2

$$A = \pi \times r^2$$

> $\pi = 3.141592.... = \underline{3.142}$ (approx)
> The big thing to remember is that π (called "pi") is just an <u>ordinary</u>
> <u>number</u> (3.14159...) which is usually rounded off to 3.142. You can also
> use the π button on your calculator (which is way more accurate).

Tangents, Chords, Arcs, Sectors and Segments (E)

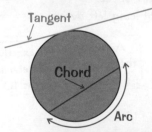

> A <u>TANGENT</u> is a straight line that <u>just touches</u> the <u>outside</u> of a circle.
> A <u>CHORD</u> is a line drawn <u>across the inside</u> of a circle.
> <u>AN ARC</u> is just <u>part of the circumference</u> of a circle.

> A <u>SECTOR</u> is a wedge-shaped area
> (like a slice of cake) cut right from the centre.
> <u>SEGMENTS</u> are the areas you get
> when you cut a circle with a chord.

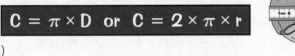

Mmm — pi...

Learn all this stuff on circles, then try this Exam Practice Question.

Q1   Identify the parts of the circle labelled A-C
      on the diagram on the right.   [3 marks] (E)

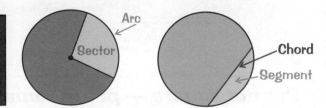

Circle Questions

ALWAYS check that you're using the right value in <u>circle formulas</u> — use the <u>radius</u> to find <u>area</u> and the <u>diameter</u> to find <u>circumference</u>. If you're given the wrong one in the question, <u>multiply</u> or <u>divide</u> by <u>2</u>.

Semicircles and Quarter Circles (C)

You might be asked to find the <u>area</u> and <u>perimeter</u> of a <u>semicircle</u> (half circle) or a <u>quarter circle</u>.

> 1) <u>AREA</u>: find the area of the <u>whole circle</u> then <u>divide</u> by <u>2</u> (for a semicircle) or <u>4</u> (for a quarter circle).
>
> 2) <u>PERIMETER</u>: divide the <u>circumference</u> by <u>2</u> (for a semicircle) or <u>4</u> (for a quarter circle) and <u>add on</u> the <u>straight edges</u> (the <u>diameter</u> for a semicircle or <u>two radiuses</u> for a quarter circle).

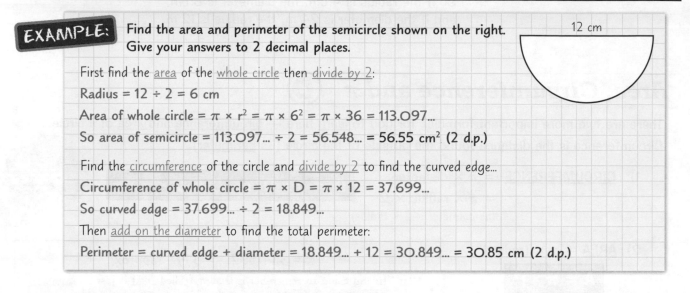

EXAMPLE: Find the area and perimeter of the semicircle shown on the right. Give your answers to 2 decimal places.

First find the <u>area</u> of the <u>whole circle</u> then <u>divide by 2</u>:

Radius = 12 ÷ 2 = 6 cm

Area of whole circle = $\pi \times r^2 = \pi \times 6^2 = \pi \times 36 = 113.097...$

So area of semicircle = 113.097... ÷ 2 = 56.548... = **56.55 cm² (2 d.p.)**

Find the <u>circumference</u> of the circle and <u>divide by 2</u> to find the curved edge...

Circumference of whole circle = $\pi \times D = \pi \times 12 = 37.699...$

So curved edge = 37.699... ÷ 2 = 18.849...

Then <u>add on the diameter</u> to find the total perimeter:

Perimeter = curved edge + diameter = 18.849... + 12 = 30.849... = **30.85 cm (2 d.p.)**

12 cm

Area Problems with Circles (C)

There's a whole range of <u>circle questions</u> you could be asked — but if you <u>learn</u> the <u>circle formulas</u> you should be fine. Make sure you <u>read</u> the questions <u>carefully</u> to find out what you're being asked to do.

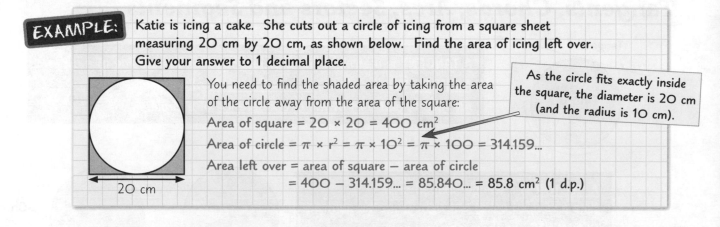

EXAMPLE: Katie is icing a cake. She cuts out a circle of icing from a square sheet measuring 20 cm by 20 cm, as shown below. Find the area of icing left over. Give your answer to 1 decimal place.

You need to find the shaded area by taking the area of the circle away from the area of the square:

Area of square = 20 × 20 = 400 cm²

Area of circle = $\pi \times r^2 = \pi \times 10^2 = \pi \times 100 = 314.159...$

Area left over = area of square − area of circle
= 400 − 314.159... = 85.840... = **85.8 cm² (1 d.p.)**

> As the circle fits exactly inside the square, the diameter is 20 cm (and the radius is 10 cm).

20 cm

Pi r not square — pi are round. Pi are tasty...

The trickiest thing about area questions is working out what you need to add on or take away to find the thing you're looking for. You can always draw a sketch to help you figure it out if you're not given one in the question.

Q1   Dave is making a tutu out of a circle of netting of diameter 120 cm.
He cuts a circular hole of diameter 30 cm out of the middle of the netting.
Find the area of netting he uses for the tutu to the nearest cm². **[3 marks]** (C)

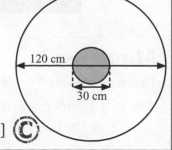

120 cm

30 cm

Volume

Remember <u>3D shapes</u>? You came across them on p60. Now it's time to work out their <u>volumes</u>.

> ### <u>LEARN</u> these volume formulas...

Volumes of Cuboids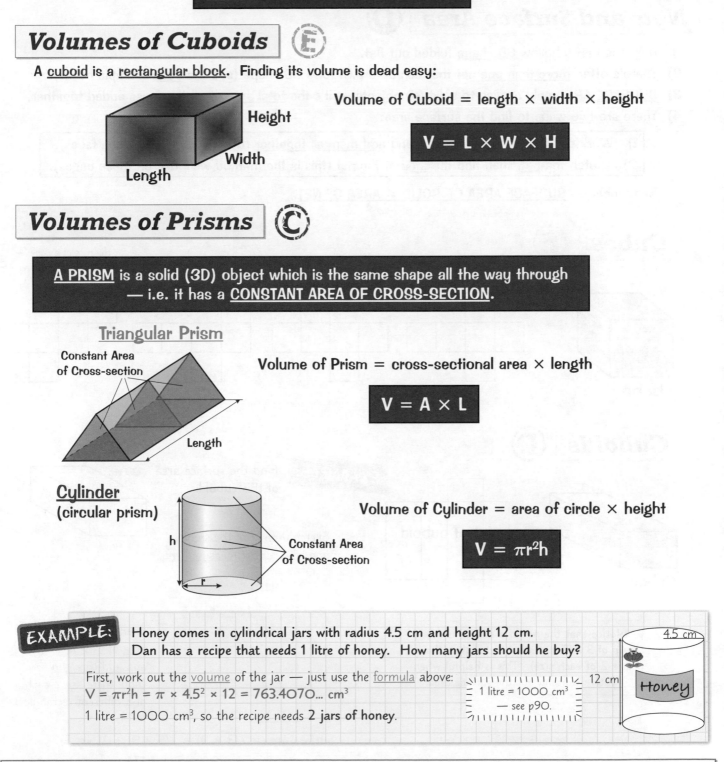

A <u>cuboid</u> is a <u>rectangular block</u>. Finding its volume is dead easy:

Height
Width
Length

Volume of Cuboid = length × width × height

$$V = L \times W \times H$$

Volumes of Prisms

> A PRISM is a solid (3D) object which is the same shape all the way through
> — i.e. it has a <u>CONSTANT AREA OF CROSS-SECTION</u>.

<u>Triangular Prism</u>

Constant Area
of Cross-section

Length

Volume of Prism = cross-sectional area × length

$$V = A \times L$$

<u>Cylinder</u>
(circular prism)

h

Constant Area
of Cross-section

r

Volume of Cylinder = area of circle × height

$$V = \pi r^2 h$$

EXAMPLE: Honey comes in cylindrical jars with radius 4.5 cm and height 12 cm.
Dan has a recipe that needs 1 litre of honey. How many jars should he buy?

First, work out the <u>volume</u> of the jar — just use the <u>formula</u> above:
$V = \pi r^2 h = \pi \times 4.5^2 \times 12 = 763.4070... \text{ cm}^3$
1 litre = 1000 cm³, so the recipe needs **2 jars of honey**.

1 litre = 1000 cm³
— see p90.

4.5 cm
12 cm
Honey

Don't make it any more angry — it's already a cross-section...

The easiest type of volume question is where you're given a shape made up of 1 cm cubes
and asked for its volume. All you have to do here is count up the cubes (not forgetting the
hidden ones at the back of the shape). Have a go at this Exam Practice Question —
you'll need your area formulas from p63.

Q1   Find the volume of the triangular prism on the right.   [3 marks]

10 cm
12 cm
6 cm

Nets and Surface Area

Pencils and rulers at the ready — you might get to do some drawing over the next two pages. Unfortunately, you're mainly limited to squares, rectangles and triangles, but you might get the odd circle to draw.

Nets and Surface Area (D)

1) A NET is just a hollow 3D shape folded out flat.
2) There's often more than one net that can be drawn for a 3D shape (see the cube example below).
3) SURFACE AREA only applies to solid 3D objects — it's the total area of all the faces added together.
4) There are two ways to find the surface area:

> 1) Work out the area of each face and add them all together (don't forget the hidden faces).
> 2) Sketch the net, then find the area of the net (this is the method we'll use on these pages).

Remember — SURFACE AREA OF SOLID = AREA OF NET.

Cubes (E)

Nets of cubes

These are just some of the nets of a cube — there are lots more.

Cube

Cuboids (D)

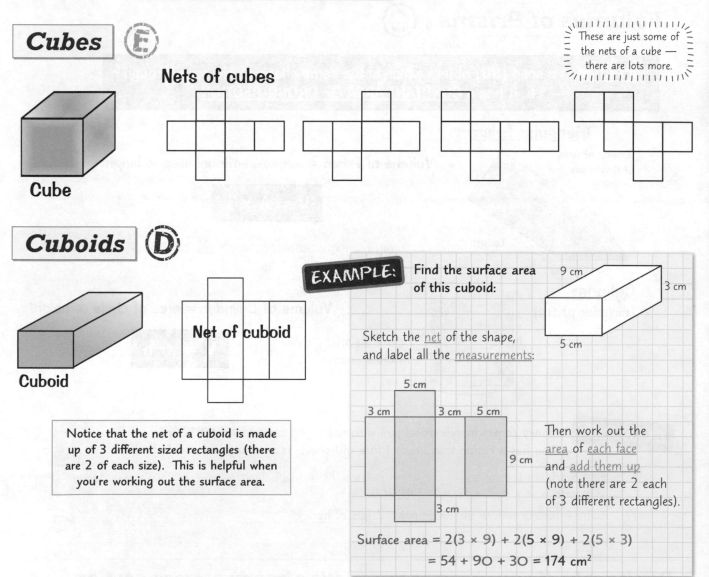

Cuboid

Net of cuboid

Notice that the net of a cuboid is made up of 3 different sized rectangles (there are 2 of each size). This is helpful when you're working out the surface area.

EXAMPLE: Find the surface area of this cuboid:

9 cm
3 cm
5 cm

Sketch the net of the shape, and label all the measurements:

5 cm
3 cm   3 cm   5 cm
9 cm
3 cm

Then work out the area of each face and add them up (note there are 2 each of 3 different rectangles).

Surface area = 2(3 × 9) + 2(5 × 9) + 2(5 × 3)
= 54 + 90 + 30 = **174 cm²**

Net yourself some extra marks...

In the exam, you might be given a number of nets and asked which ones work and which don't.
Try and imagine folding up the net in your head, and see if there are any bits that overlap or don't join up.

Q1   Draw an accurate net for a cube with side length 1.5 cm.   [2 marks] (D)

Nets and Surface Area

Another page on <u>nets</u> and <u>surface area</u> — and now things get really exciting. It's time for <u>pyramids</u>, <u>prisms</u> and <u>cylinders</u>. Ooooo.

Triangular Prisms and Pyramids Ⓒ

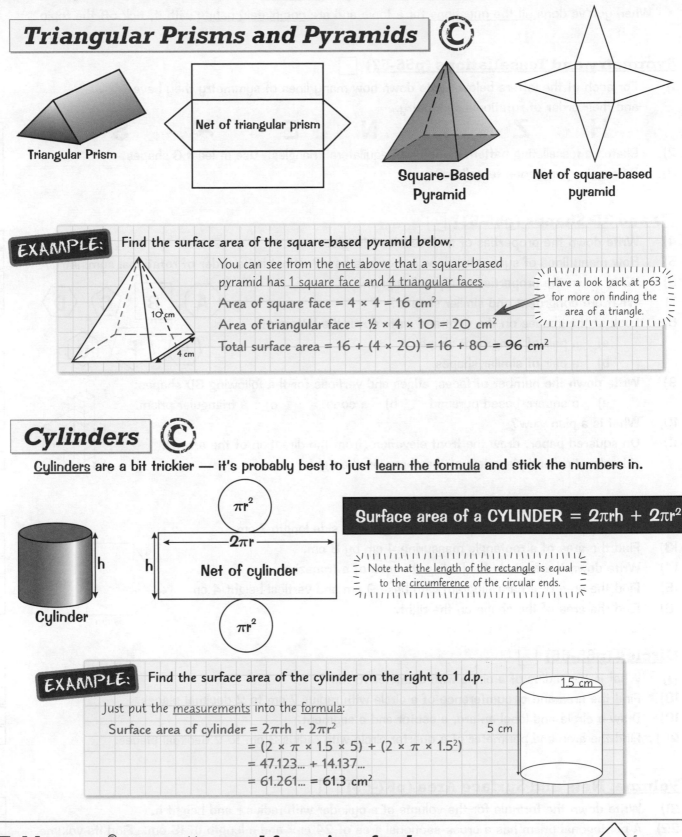

Triangular Prism

Net of triangular prism

Square-Based Pyramid

Net of square-based pyramid

EXAMPLE: Find the surface area of the square-based pyramid below.

You can see from the <u>net</u> above that a square-based pyramid has <u>1 square face</u> and <u>4 triangular faces</u>.

Area of square face = 4 × 4 = 16 cm²

Area of triangular face = ½ × 4 × 10 = 20 cm²

Total surface area = 16 + (4 × 20) = 16 + 80 = **96 cm²**

10 cm

4 cm

Have a look back at p63 for more on finding the area of a triangle.

Cylinders Ⓒ

<u>Cylinders</u> are a bit trickier — it's probably best to just <u>learn the formula</u> and stick the numbers in.

πr^2

$2\pi r$

Net of cylinder

πr^2

h

h

Cylinder

Surface area of a CYLINDER = $2\pi rh + 2\pi r^2$

Note that <u>the length of the rectangle</u> is equal to the <u>circumference</u> of the circular ends.

EXAMPLE: Find the surface area of the cylinder on the right to 1 d.p.

Just put the <u>measurements</u> into the <u>formula</u>:

Surface area of cylinder = $2\pi rh + 2\pi r^2$

= (2 × π × 1.5 × 5) + (2 × π × 1.5²)

= 47.123... + 14.137...

= 61.261... = **61.3 cm²**

1.5 cm

5 cm

Toblerones® — pretty good for prism food...

Make sure you know the formulas for finding the area of triangles and circles, then have a go at this Exam Practice Question.

Q1   Find the surface area of this triangular prism.   [3 marks] Ⓒ

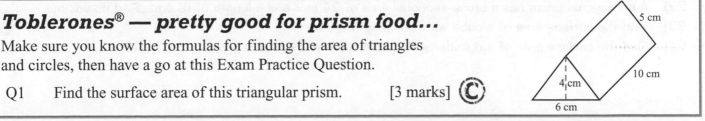

5 cm

10 cm

4 cm

6 cm

Revision Questions for Section Four

Lots of lovely shapes and formulas to learn in Section 4 — now it's time to see what's sunk in.

- Try these questions and <u>tick off each one</u> when you <u>get it right</u>.
- When you've done <u>all the questions</u> for a topic and are <u>completely happy</u> with it, tick off the topic.

Symmetry and Tessellations (p56-57) ☑

1) For each of the letters below, write down how many lines of symmetry they have and their order of rotational symmetry.

H   Z   T   N   E   ✖   S

2) Sketch a tessellating pattern made up of equilateral triangles. Use at least 6 shapes.

3) Do regular pentagons tessellate?

2D and 3D Shapes (p58-61) ☑

4) Write down the properties of an isosceles triangle.

5) How many lines of symmetry does a rhombus have? What is its order of rotational symmetry?

6) Write down the properties of a parallelogram.

7) What are congruent and similar shapes?

8) Look at the shapes on the right and write down the letters of:

   a) a pair of congruent shapes,

   b) a pair of similar shapes.

9) Write down the number of faces, edges and vertices for the following 3D shapes:

   a) a square-based pyramid   b) a cone   c) a triangular prism.

10) What is a plan view?

11) On squared paper, draw the front elevation (from the direction of the arrow), side elevation and plan view of the shape on the right.

Perimeter and Area (p62-64) ☑

12) Find the perimeter of an equilateral triangle with side length 7 cm.

13) Find the area of a rectangle measuring 4 cm by 8 cm.

14) Write down the formula for finding the area of a trapezium.

15) Find the area of a parallelogram with base 9 cm and vertical height 4 cm.

16) Find the area of the shape on the right.

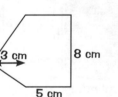

Circles (p65-66) ☑

17) What is the radius of a circle with diameter 18 mm?

18) Find the area and circumference of a circle with radius 7 cm to 2 decimal places.

19) Draw a circle and label an arc, a sector and a segment.

20) Find the area and perimeter of a quarter circle with radius 3 cm to 2 decimal places.

Volume, Nets and Surface Area (p67-69) ☑

21) Write down the formula for the volume of a cylinder with radius r and height h.

22) A pentagonal prism has a cross-sectional area of 24 cm² and a length of 15 cm. Find its volume.

23) Find the surface area of a cube with side length 5 cm.

24) Find the surface area of a cylinder with height 8 cm and radius 2 cm to 1 d.p.

Lines and Angles

Before we really get going with the thrills and chills of angles and geometry, there are a few things you need to know. Nothing too scary — just some <u>special angles</u> and some <u>fancy notation</u>.

Four Special Angles Ⓖ

There are <u>360°</u> in a full turn, and it can be divided into 4 special angles:

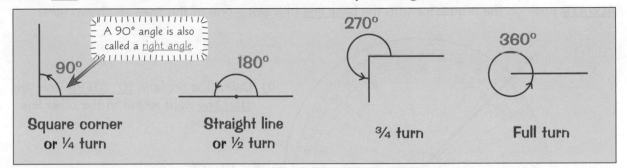

A 90° angle is also called a <u>right angle</u>.

90°	180°	270°	360°
Square corner or ¼ turn	Straight line or ½ turn	¾ turn	Full turn

When two lines meet at 90° they are said to be <u>PERPENDICULAR</u> to each other.

Fancy Angle Names Ⓕ

Some angles have special names. You might have to <u>identify</u> these angles in the exam.

ACUTE angles
Sharp pointy ones
(less than 90°)

RIGHT angles
Square corners
(exactly 90°)

OBTUSE angles
Flatter ones
(between 90° and 180°)

REFLEX angles
Ones that bend back on themselves
(more than 180°)

Three-Letter Angle Notation Ⓕ

The best way to say which angle you're talking about in a diagram is by using <u>THREE</u> letters.
For example in the diagram, angle ACB = 25°.

1) The <u>middle letter</u> is where the angle is.
2) The <u>other two letters</u> tell you which two lines enclose the angle.

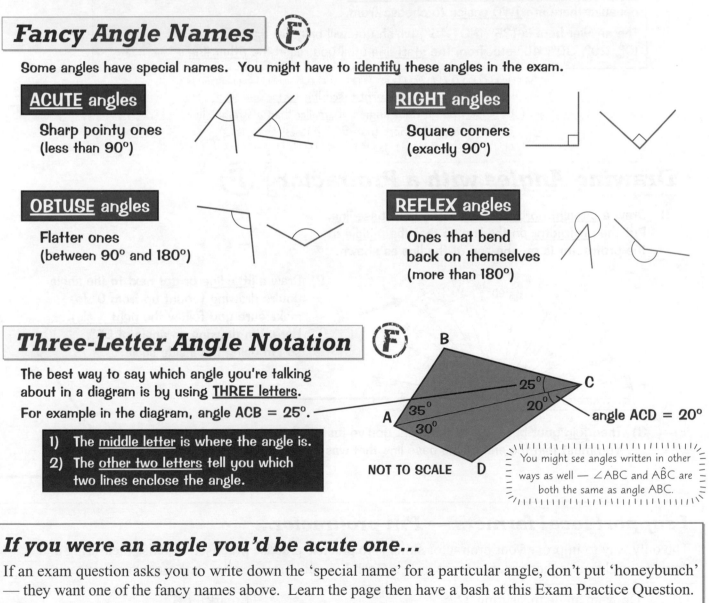

angle ACD = 20°

NOT TO SCALE

You might see angles written in other ways as well — ∠ABC and AB̂C are both the same as angle ABC.

If you were an angle you'd be acute one...

If an exam question asks you to write down the 'special name' for a particular angle, don't put 'honeybunch' — they want one of the fancy names above. Learn the page then have a bash at this Exam Practice Question.

Q1   a) An angle measures 66°. What type of angle is this?   [1 mark]
       b) Write down an example of a reflex angle.   [1 mark] Ⓕ

Measuring & Drawing Angles

The <u>2 big mistakes</u> that people make with protractors:

> 1) Not putting the <u>0° line</u> at the <u>start</u> position.
> 2) Reading from the <u>WRONG SCALE</u>.

Measuring Angles with a Protractor (F)

1) <u>ALWAYS</u> position the protractor with the <u>base line</u> of it along one of the lines as shown here:

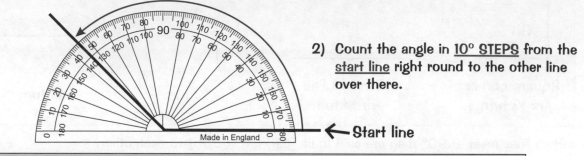

2) Count the angle in <u>10° STEPS</u> from the <u>start line</u> right round to the other line over there.

← Start line

<u>DON'T JUST READ A NUMBER OFF THE SCALE</u> — chances are it'll be the wrong one because there are <u>TWO scales</u> to choose from.

The answer here is 135° (NOT 45°) which you will only get right if you start counting 10°, 20°, 30°, 40° etc. from the <u>start line</u> until you reach the <u>other line</u>.

> <u>Check your measurement is sensible by looking at it</u>
> If it's <u>bigger</u> than a right angle but <u>smaller</u> than a straight line,
> then it must be more than 90° but less than 180°.

Drawing Angles with a Protractor (F)

1) Draw a <u>straight horizontal line</u> to be your <u>base line</u>. Put the <u>protractor</u> on the line so that the <u>middle</u> of the protractor is on one <u>end</u> of the line as shown:

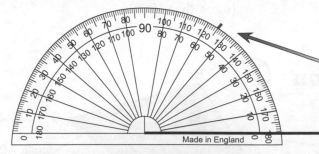

2) Draw a <u>little line</u> or <u>dot</u> next to the angle you're drawing (count up from 0° to make sure you follow the <u>right scale</u>). Here, I'm drawing an angle of 55°, so I'm using the <u>inside</u> scale.

3) Then <u>join</u> your <u>base line</u> to the <u>mark</u> you've just made with a <u>straight line</u>. You must join the end of the base line that was in the <u>middle</u> of the protractor.

55°

I support local farmers — I'm pro-tractor...

The only way to improve your protractor skills is to practise practise practise. Practise measuring and drawing angles, and then practise some more. Gosh I've said practise a lot in those two sentences.

Q1   Draw an angle measuring 227°.   [1 mark] (F)

Q2   Measure the angle on the right.   [1 mark] (F)

Five Angle Rules

If you know <u>all</u> these rules <u>thoroughly</u>, you'll at least have a fighting chance of working out problems with lines and angles. If you don't — you've no chance. Sorry to break it to you like that.

5 Simple Rules — That's All

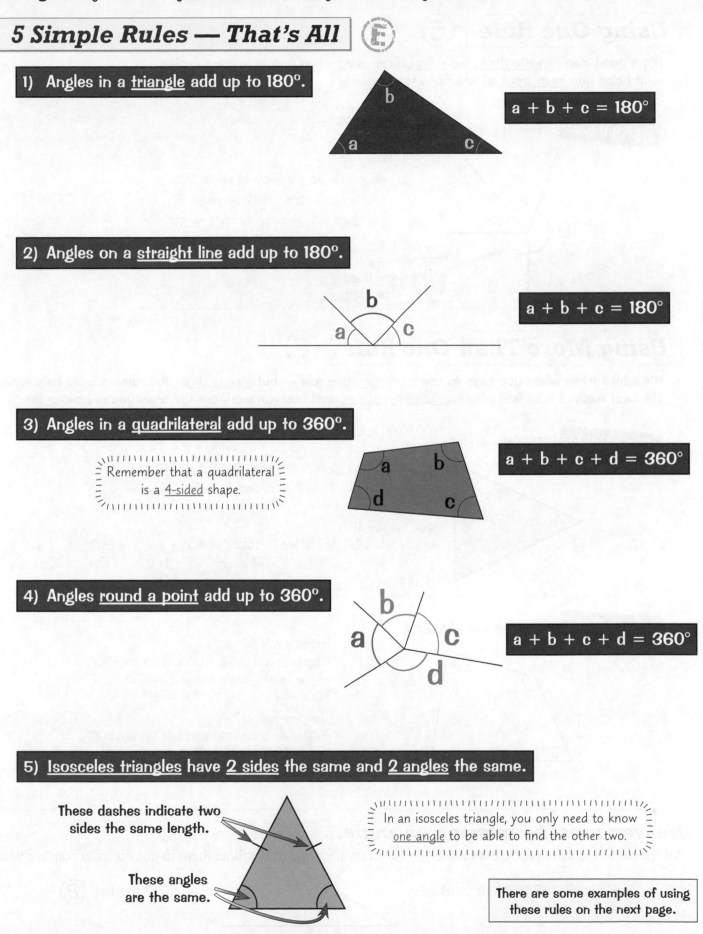

1) Angles in a <u>triangle</u> add up to 180°.

$$a + b + c = 180°$$

2) Angles on a <u>straight line</u> add up to 180°.

$$a + b + c = 180°$$

3) Angles in a <u>quadrilateral</u> add up to 360°.

Remember that a quadrilateral is a <u>4-sided</u> shape.

$$a + b + c + d = 360°$$

4) Angles <u>round a point</u> add up to 360°.

$$a + b + c + d = 360°$$

5) <u>Isosceles triangles</u> have <u>2 sides</u> the same and <u>2 angles</u> the same.

These dashes indicate two sides the same length.

In an isosceles triangle, you only need to know <u>one angle</u> to be able to find the other two.

These angles are the same.

There are some examples of using these rules on the next page.

Section Five — Angles and Geometry

Five Angle Rules

Right, by now you should know the <u>five angle rules</u> (if you're not sure about them, go back over the previous page until you know them really well). Now it's time to see them <u>in action</u>.

Using One Rule (E)

It's a good idea to <u>write down</u> the <u>rules</u> you're using when finding missing angles — it helps you <u>keep track</u> of what you're doing.

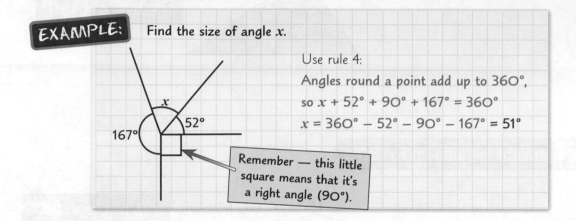

EXAMPLE: Find the size of angle x.

167°

x

52°

Use rule 4:
Angles round a point add up to 360°,
so x + 52° + 90° + 167° = 360°
x = 360° − 52° − 90° − 167° = 51°

Remember — this little square means that it's a right angle (90°).

Using More Than One Rule (E)

It's a bit trickier when you have to use <u>more than one</u> rule — but <u>writing down</u> the rules is a big help again. The best method is to find <u>whatever angles you can</u> until you can work out the ones you're looking for.

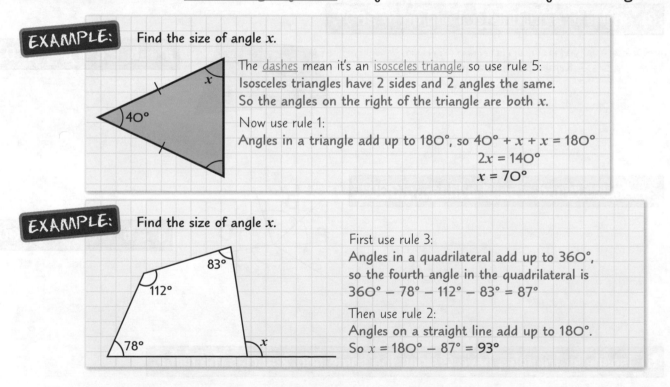

EXAMPLE: Find the size of angle x.

40°

x

The <u>dashes</u> mean it's an <u>isosceles triangle</u>, so use rule 5:
Isosceles triangles have 2 sides and 2 angles the same.
So the angles on the right of the triangle are both x.

Now use rule 1:
Angles in a triangle add up to 180°, so 40° + x + x = 180°
2x = 140°
x = 70°

EXAMPLE: Find the size of angle x.

83°

112°

78°

x

First use rule 3:
Angles in a quadrilateral add up to 360°,
so the fourth angle in the quadrilateral is
360° − 78° − 112° − 83° = 87°

Then use rule 2:
Angles on a straight line add up to 180°.
So x = 180° − 87° = 93°

Heaven must be missing an angle...

All the basic facts are pretty easy really — but examiners like to combine them in questions to confuse you.

Q1 Find the size of the angle marked x.

72°

x

[2 marks] (E)

Parallel Lines

Parallel lines are always the <u>same distance apart</u>. This page is all about them.

Angles Around Parallel Lines Ⓓ

> You also need
> to know what
> <u>perpendicular lines</u> are
> — they meet at <u>90°</u>.

When a line crosses two <u>parallel lines</u>...

> 1) The two bunches of angles are <u>the same</u>.
> 2) There are <u>only two different angles</u>: <u>a small one</u> and <u>a big one</u>.
> 3) These <u>ALWAYS ADD UP TO 180°</u>. E.g. 30° and 150° below

The two lines with the <u>arrows</u> on are <u>parallel</u>:

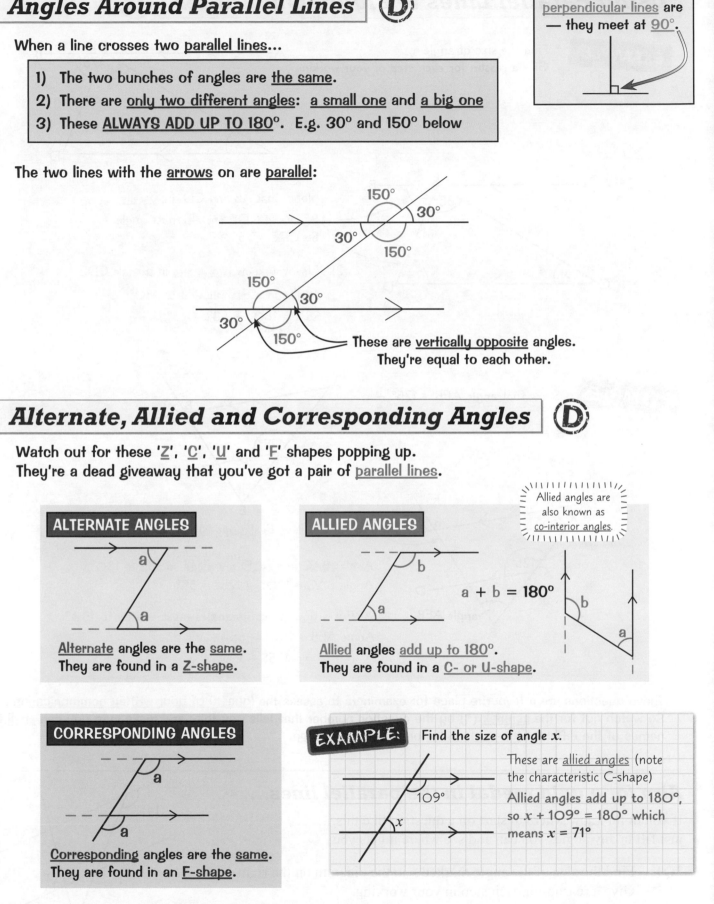

These are <u>vertically opposite</u> angles.
They're equal to each other.

Alternate, Allied and Corresponding Angles Ⓓ

Watch out for these '<u>Z</u>', '<u>C</u>', '<u>U</u>' and '<u>F</u>' shapes popping up.
They're a dead giveaway that you've got a pair of <u>parallel lines</u>.

ALTERNATE ANGLES

<u>Alternate</u> angles are the <u>same</u>.
They are found in a <u>Z-shape</u>.

ALLIED ANGLES

> Allied angles are
> also known as
> <u>co-interior angles</u>.

$a + b = 180°$

<u>Allied</u> angles <u>add up to 180°</u>.
They are found in a <u>C- or U-shape</u>.

CORRESPONDING ANGLES

<u>Corresponding</u> angles are the <u>same</u>.
They are found in an <u>F-shape</u>.

EXAMPLE: Find the size of angle x.

These are <u>allied angles</u> (note
the characteristic C-shape)

Allied angles add up to 180°,
so $x + 109° = 180°$ which
means $x = 71°$

Section Five — Angles and Geometry

Parallel Lines

Exam questions involving parallel lines can be pretty involved — that's why there are some big examples below.

Using Parallel Lines to Figure Out Angles Ⓓ

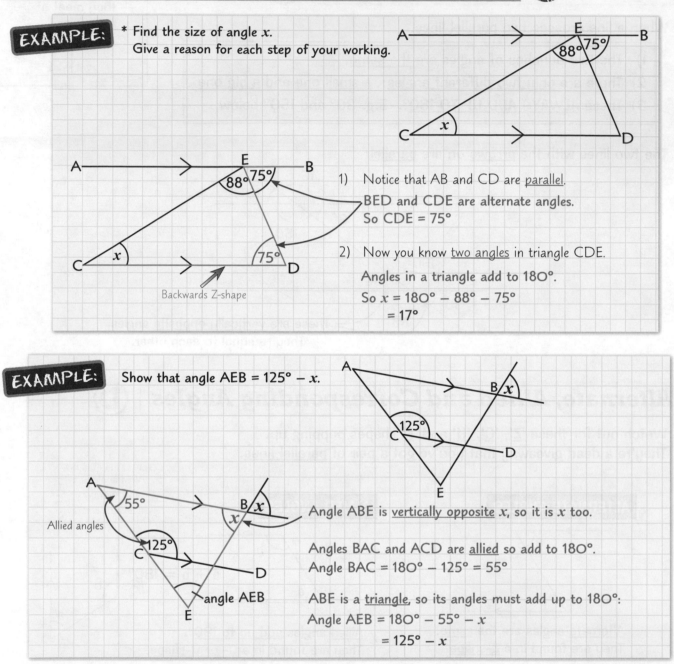

EXAMPLE:
* Find the size of angle x.
Give a reason for each step of your working.

1) Notice that AB and CD are parallel.
BED and CDE are alternate angles.
So CDE = 75°

2) Now you know two angles in triangle CDE.
Angles in a triangle add to 180°.
So x = 180° − 88° − 75°
    = 17°

Backwards Z-shape

EXAMPLE: Show that angle AEB = 125° − x.

Allied angles

angle AEB

Angle ABE is vertically opposite x, so it is x too.

Angles BAC and ACD are allied so add to 180°.
Angle BAC = 180° − 125° = 55°

ABE is a triangle, so its angles must add up to 180°:
Angle AEB = 180° − 55° − x
       = 125° − x

These questions are a favourite place for examiners to assess the 'quality of your written communication'.
So watch out for the asterisk (*) by the question number that tells you this, and make sure you can spell the
names of the different types of angles on the previous page.

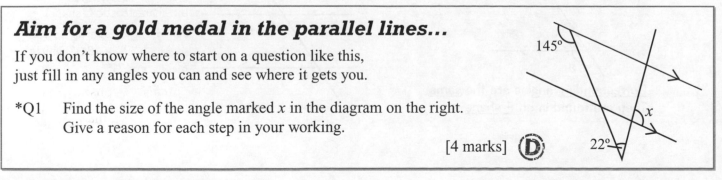

Aim for a gold medal in the parallel lines...

If you don't know where to start on a question like this,
just fill in any angles you can and see where it gets you.

*Q1 Find the size of the angle marked x in the diagram on the right.
Give a reason for each step in your working.

[4 marks] Ⓓ

Polygons and Angles

A polygon is a <u>many-sided shape</u>, and can be <u>regular</u> or <u>irregular</u>. A regular polygon is one where all the sides and angles are the <u>same</u> (in an irregular polygon, the sides and angles are <u>different</u>).

Regular Polygons   Ⓔ

Learn the names of these <u>regular polygons</u> and how many <u>sides</u> they have (remember that all the <u>sides</u> and <u>angles</u> in a regular polygon are the <u>same</u>).

<u>EQUILATERAL TRIANGLE</u>
3 sides

<u>SQUARE</u>
(regular quadrilateral)
4 sides

<u>PENTAGON</u>
5 sides

<u>HEXAGON</u>
6 sides

<u>HEPTAGON</u>
7 sides

<u>OCTAGON</u>
8 sides

<u>DECAGON</u>
10 sides

Exterior and Interior Angles   Ⓓ

You need to know <u>what</u> exterior and interior angles are and <u>how to find them</u>.

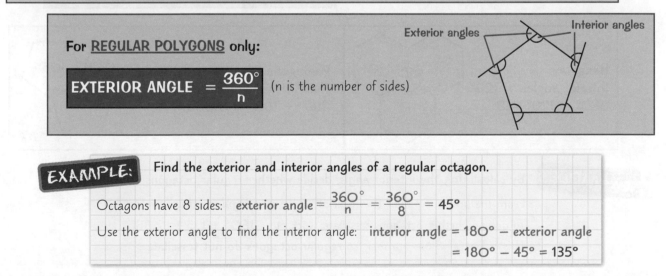

For <u>ANY POLYGON</u> (regular or irregular):

SUM OF EXTERIOR ANGLES = 360°

INTERIOR ANGLE = 180° – EXTERIOR ANGLE

Exterior angle

Interior angle

For <u>REGULAR POLYGONS</u> only:

EXTERIOR ANGLE $= \dfrac{360°}{n}$ (n is the number of sides)

Exterior angles

Interior angles

EXAMPLE: Find the exterior and interior angles of a regular octagon.

Octagons have 8 sides: exterior angle $= \dfrac{360°}{n} = \dfrac{360°}{8} = 45°$

Use the exterior angle to find the interior angle: interior angle $= 180° -$ exterior angle
$= 180° - 45° = 135°$

EXCLUSIVE: Heptagon lottery winner says "I'm still just a regular guy"...

Learn all the formulas above, and remember whether they go with regular or irregular polygons.

Q1   Find the size of the interior angle of a regular decagon. [2 marks] Ⓓ

Q2   A regular polygon has exterior angles of 72°. What is the name of the polygon? [2 marks] Ⓓ

Polygons, Angles and Tessellations

Just one more <u>polygon angle formula</u>. Then I'll be unveiling <u>why some polygons tessellate</u> and others don't.

The Tricky One — Sum of Interior Angles Ⓓ

This formula for the <u>sum of the interior angles</u> works for <u>ALL</u> polygons, even irregular ones.

SUM OF INTERIOR ANGLES = (n – 2) × 180° (n is the number of sides)

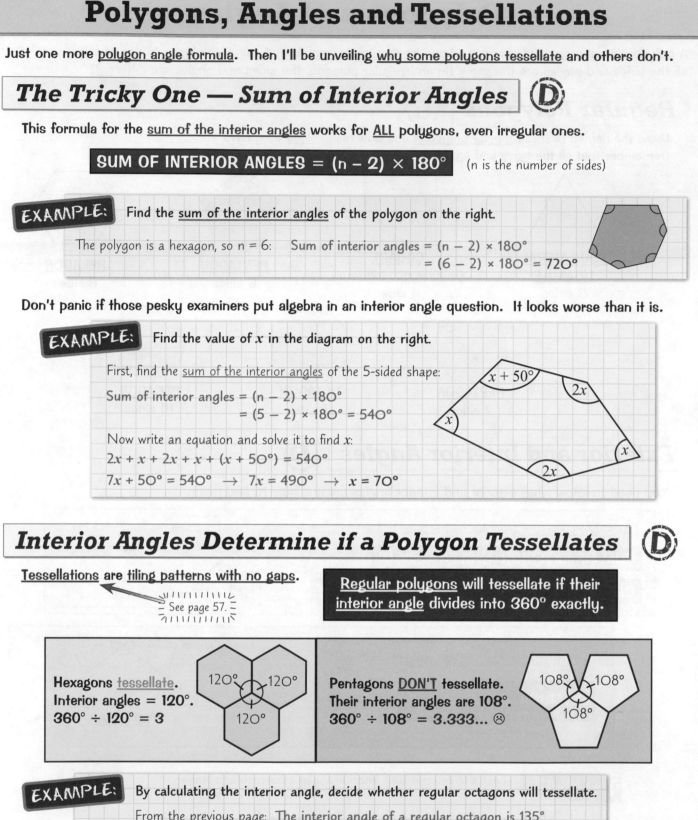

EXAMPLE: Find the <u>sum of the interior angles</u> of the polygon on the right.

The polygon is a hexagon, so n = 6:   Sum of interior angles = (n – 2) × 180°
= (6 – 2) × 180° = 720°

Don't panic if those pesky examiners put algebra in an interior angle question. It looks worse than it is.

EXAMPLE: Find the value of x in the diagram on the right.

First, find the <u>sum of the interior angles</u> of the 5-sided shape:

Sum of interior angles = (n – 2) × 180°
= (5 – 2) × 180° = 540°

Now write an equation and solve it to find x:

$2x + x + 2x + x + (x + 50°) = 540°$

$7x + 50° = 540°$ → $7x = 490°$ → $x = 70°$

Diagram labels: $x + 50°$, $2x$, x, x, $2x$

Interior Angles Determine if a Polygon Tessellates Ⓓ

<u>Tessellations</u> are <u>tiling patterns with no gaps</u>.
— See page 57. —

Regular polygons will tessellate if their **interior angle** divides into 360° exactly.

Hexagons <u>tessellate</u>.
Interior angles = 120°.
360° ÷ 120° = 3

Diagram labels: 120°, 120°, 120°

Pentagons <u>DON'T</u> tessellate.
Their interior angles are 108°.
360° ÷ 108° = 3.333... ☹

Diagram labels: 108°, 108°, 108°

EXAMPLE: By calculating the interior angle, decide whether regular octagons will tessellate.

From the previous page: The interior angle of a regular octagon is 135°

To decide whether octagons tessellate, work out: 360° ÷ 135° = 2.666...

This is not a whole number, so **regular octagons do not tessellate**.

I'm not going to make the obvious joke. We're both above that...

So to find out if shapes will form a tiling pattern with no gaps, try to fit them around a point.

Q1   a) Find the sum of the interior angles of a regular heptagon.   [2 marks] Ⓓ

b) Show that regular heptagons do not tessellate.   [2 marks] Ⓓ

Aww man, this was gonna be my big break an' everythin'.

Transformations

There are four <u>transformations</u> you need to know — <u>translation</u>, <u>reflection</u>, <u>rotation</u> and <u>enlargement</u>.

1) Translations D

A translation is just a <u>SLIDE</u> around the page. When describing a translation, you must say <u>how far along</u> and <u>how far up</u> the shape moves using a vector.

> <u>Vectors</u> describing translations look like this. ⟶
> x is the number of spaces <u>right</u>, y is the number of spaces <u>up</u>. $\begin{pmatrix} x \to \\ \uparrow y \end{pmatrix}$

If the shape moves <u>left</u> x will be <u>negative</u>, and if it moves <u>down</u> y will be <u>negative</u>.

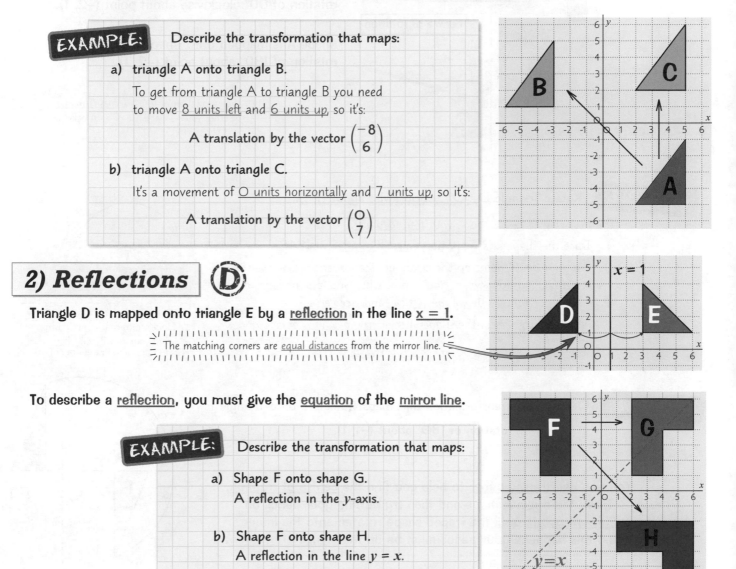

EXAMPLE: Describe the transformation that maps:

a) triangle A onto triangle B.

To get from triangle A to triangle B you need to move <u>8 units left</u> and <u>6 units up</u>, so it's:

A translation by the vector $\begin{pmatrix} -8 \\ 6 \end{pmatrix}$

b) triangle A onto triangle C.

It's a movement of <u>0 units horizontally</u> and <u>7 units up</u>, so it's:

A translation by the vector $\begin{pmatrix} 0 \\ 7 \end{pmatrix}$

2) Reflections D

Triangle D is mapped onto triangle E by a <u>reflection</u> in the line <u>x = 1</u>.

> The matching corners are <u>equal distances</u> from the mirror line.

To describe a <u>reflection</u>, you must give the <u>equation</u> of the <u>mirror line</u>.

EXAMPLE: Describe the transformation that maps:

a) Shape F onto shape G.
A reflection in the *y*-axis.

b) Shape F onto shape H.
A reflection in the line *y* = *x*.

Moving eet to ze left — a perfect translation...

You're allowed to use tracing paper in the exam — use it to check your answers to reflection questions. Trace the original shape and the mirror line, then flip the tracing paper over and line up the mirror lines. If you've done the reflection correctly, the shapes will match up perfectly.

Q1   Describe the transformation that maps triangle C onto triangle B (above).   [2 marks] D

Q2   On a grid, copy triangle A above and reflect it in the line *y* = 0.   [1 mark] D

More on Transformations

Transformation number 3 coming up. Rotation.

3) Rotations Ⓓ

To describe a rotation, you need 3 details:

1) The angle of rotation (usually 90° or 180°).
2) The direction of rotation (clockwise or anticlockwise).
3) The centre of rotation

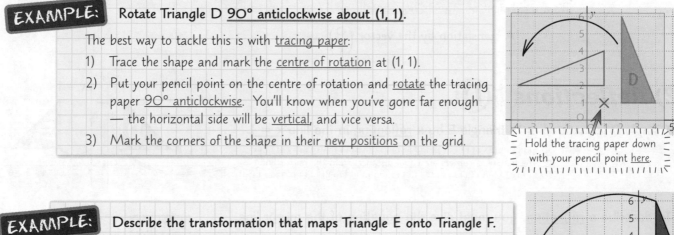

centre of rotation

Shape A is mapped onto Shape B by a rotation of 90° clockwise about point (−2, 1).

Shape A is mapped onto Shape C by a rotation of 180° about point (−2, 1).

For a rotation of 180°, it doesn't matter whether you go clockwise or anticlockwise.

EXAMPLE: Rotate Triangle D 90° anticlockwise about (1, 1).

The best way to tackle this is with tracing paper:

1) Trace the shape and mark the centre of rotation at (1, 1).

2) Put your pencil point on the centre of rotation and rotate the tracing paper 90° anticlockwise. You'll know when you've gone far enough — the horizontal side will be vertical, and vice versa.

3) Mark the corners of the shape in their new positions on the grid.

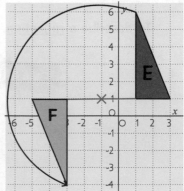

Hold the tracing paper down with your pencil point here.

EXAMPLE: Describe the transformation that maps Triangle E onto Triangle F.
A rotation of 180° about (−1, 1).

You can use tracing paper to help you find the centre of rotation. Trace the original shape and then try putting your pencil on different points until the traced shape rotates onto the image. When this happens your pencil must be on the centre of rotation.

Tracing paper — thank goodness school loo roll moved on...

That's all the transformations which don't change the size of a shape.
Try this question about translation, reflection and rotation.

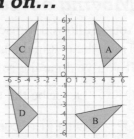

Q1   Describe the transformations which map:
a) A onto B   [3 marks]   b) B onto C   [2 marks]
c) C onto A   [2 marks]   d) A onto D   [2 marks] Ⓓ

Transformations — Enlargements

One more transformation coming up — <u>enlargements</u>. They're the trickiest, but also the most fun (honest).

4) Enlargements Ⓓ

The <u>scale factor</u> for an enlargement tells you <u>how long</u> the sides of the new shape are compared to the old shape. E.g. a scale factor of 3 means you <u>multiply</u> each side length by 3.

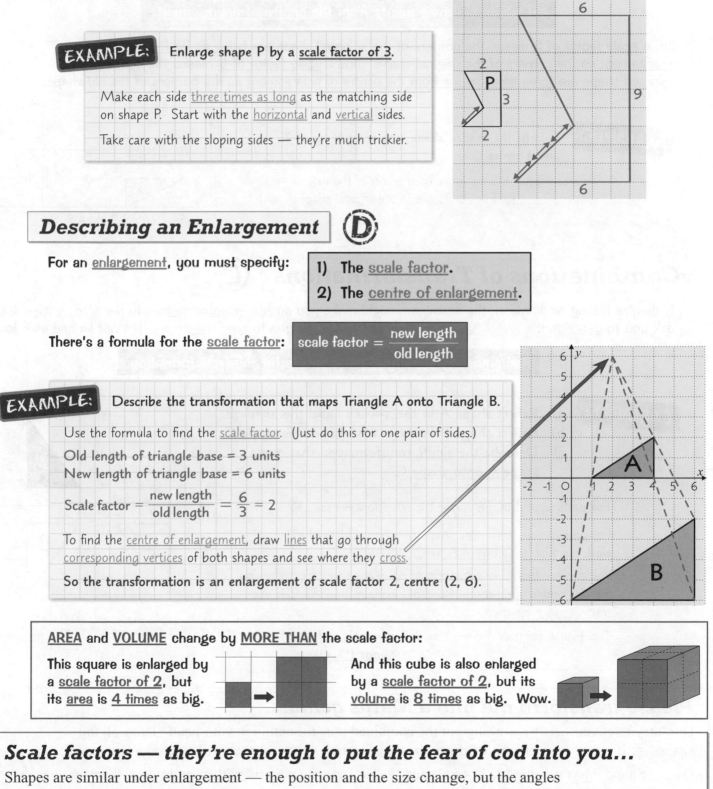

EXAMPLE: Enlarge shape P by a <u>scale factor of 3</u>.

Make each side <u>three times as long</u> as the matching side on shape P. Start with the <u>horizontal</u> and <u>vertical</u> sides.

Take care with the sloping sides — they're much trickier.

Describing an Enlargement Ⓓ

For an <u>enlargement</u>, you must specify:

1) The <u>scale factor</u>.
2) The <u>centre of enlargement</u>.

There's a formula for the <u>scale factor</u>: $$\text{scale factor} = \frac{\text{new length}}{\text{old length}}$$

EXAMPLE: Describe the transformation that maps Triangle A onto Triangle B.

Use the formula to find the <u>scale factor</u>. (Just do this for one pair of sides.)

Old length of triangle base = 3 units
New length of triangle base = 6 units

Scale factor = $\dfrac{\text{new length}}{\text{old length}} = \dfrac{6}{3} = 2$

To find the <u>centre of enlargement</u>, draw <u>lines</u> that go through <u>corresponding vertices</u> of both shapes and see where they <u>cross</u>.

So the transformation is an enlargement of scale factor 2, centre (2, 6).

<u>AREA</u> and <u>VOLUME</u> change by <u>MORE THAN</u> the scale factor:

This square is enlarged by a <u>scale factor of 2</u>, but its <u>area</u> is <u>4 times</u> as big.

And this cube is also enlarged by a <u>scale factor of 2</u>, but its <u>volume</u> is <u>8 times</u> as big. Wow.

Scale factors — they're enough to put the fear of cod into you...

Shapes are similar under enlargement — the position and the size change, but the angles and ratios of the sides don't (see p59).

Q1   On a grid, draw triangle A with vertices (2, 1), (4, 1) and (4, 3), and triangle B with vertices (3, 1), (7, 1) and (7, 5). Describe the transformation that maps A to B.   [4 marks] Ⓓ

Harder Transformations

Just one more page on transformations, and then you're done. With transformations anyway, not with Maths.

Enlarging a Shape with a Given Centre of Enlargement (C)

If you're given the <u>centre of enlargement</u>, then it's vitally important <u>where</u> your new shape is on the grid.

> The <u>scale factor</u> tells you the **RELATIVE DISTANCE** of the old points and new points from the <u>centre of enlargement</u>.

So, a <u>scale factor of 2</u> means the corners of the enlarged shape are <u>twice as far from the centre of enlargement</u> as the corners of the original shape. And a <u>scale factor of 3</u> means the corners of the enlarged shape are <u>three times as far from the centre of enlargement</u> as the corners of the old shape.

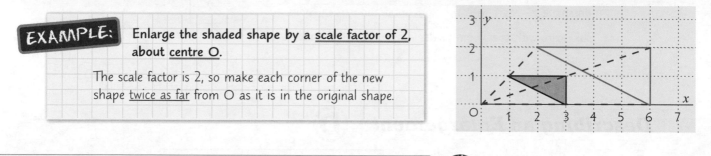

EXAMPLE: Enlarge the shaded shape by a <u>scale factor of 2</u>, about <u>centre O</u>.

The scale factor is 2, so make each corner of the new shape <u>twice as far</u> from O as it is in the original shape.

Combinations of Transformations (C)

If they're feeling really mean, the examiners might make you do <u>two transformations</u> to the <u>same shape</u>, then ask you to <u>describe</u> the <u>single transformation</u> that would get you to the <u>final shape</u>. It's not as bad as it looks.

> Remember to specify **ALL** the details for the transformation.

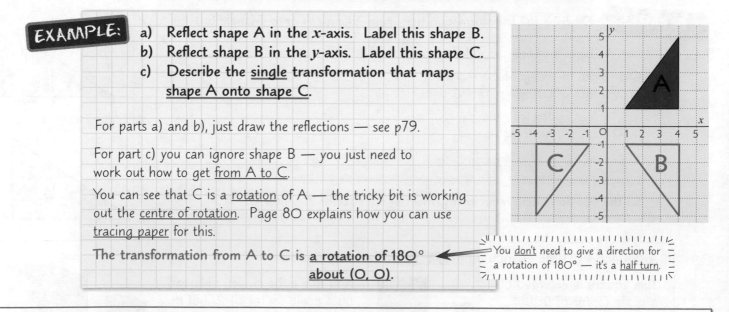

EXAMPLE:
a) Reflect shape A in the *x*-axis. Label this shape B.
b) Reflect shape B in the *y*-axis. Label this shape C.
c) Describe the <u>single</u> transformation that maps <u>shape A onto shape C</u>.

For parts a) and b), just draw the reflections — see p79.

For part c) you can ignore shape B — you just need to work out how to get <u>from A to C</u>.

You can see that C is a <u>rotation</u> of A — the tricky bit is working out the <u>centre of rotation</u>. Page 80 explains how you can use <u>tracing paper</u> for this.

The transformation from A to C is <u>a rotation of 180°</u> <u>about (0, 0)</u>.

You <u>don't</u> need to give a direction for a rotation of 180° — it's a <u>half turn</u>.

Please transform me into a Maths genius...

Don't try to cut corners by working out the combined transformation in your head. Draw all the shapes — if you slip up on writing down the transformation these shapes will get you a few marks.

Q1   Shape P has vertices (2, 2), (5, 2) and (5, 4). Draw it on a grid with *x*- and *y*-axes from –6 to 6. (C)

    a)   Shape P is rotated 180° about (2, 1) to give shape Q. Draw shape Q on the grid.   [3 marks]

    b)   Shape Q is then translated $\begin{pmatrix} 3 \\ -2 \end{pmatrix}$ to give shape R. Draw shape R on the grid.   [3 marks]

    c)   Describe the single transformation that will map shape R onto P.   [3 marks]

Similar Shape Problems

Similar shapes are <u>exactly the same shape</u>, but are <u>different sizes</u> (they can also be <u>rotated</u> or <u>reflected</u>).

Similar Shapes Have the Same Angles Ⓔ

Two shapes are <u>similar</u> if:

1) All the <u>angles</u> match up.

2) The <u>sides</u> are all enlarged by the <u>same scale factor</u>.

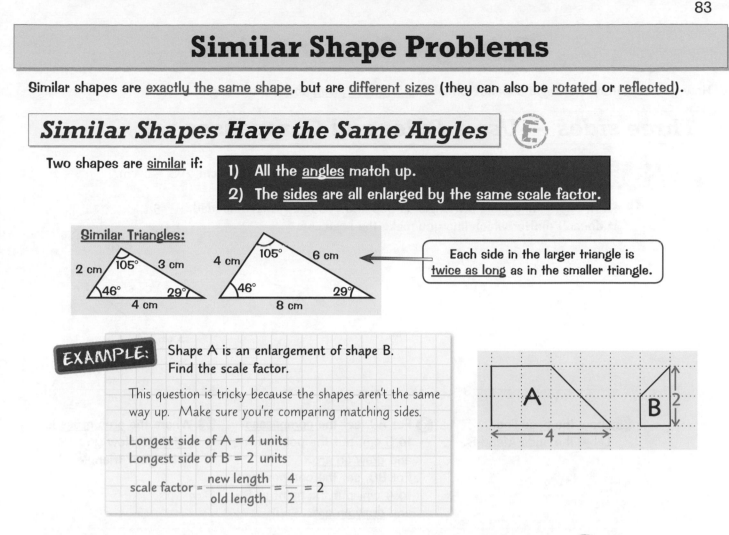

<u>Similar Triangles:</u>

2 cm | 105° | 3 cm | 4 cm | 105° | 6 cm
46° | 29° | 4 cm | 46° | 29° | 8 cm

Each side in the larger triangle is <u>twice as long</u> as in the smaller triangle.

EXAMPLE: Shape A is an enlargement of shape B. Find the scale factor.

This question is tricky because the shapes aren't the same way up. Make sure you're comparing matching sides.

Longest side of A = 4 units
Longest side of B = 2 units

scale factor = $\dfrac{\text{new length}}{\text{old length}} = \dfrac{4}{2} = 2$

Use the Scale Factor to Find Missing Sides Ⓔ

Exam questions often ask you to find the <u>length</u> of a <u>missing side</u> or the <u>size</u> of a <u>missing angle</u> in a pair of similar shapes.

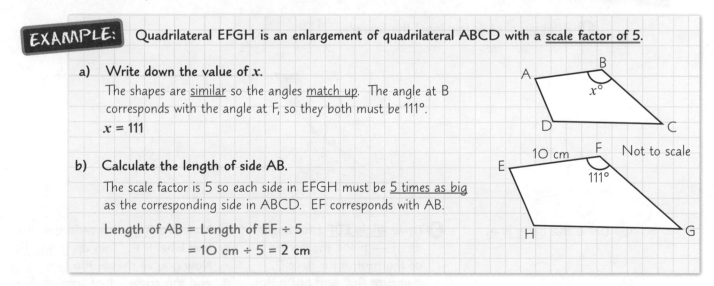

EXAMPLE: Quadrilateral EFGH is an enlargement of quadrilateral ABCD with a <u>scale factor of 5</u>.

a) Write down the value of x.

The shapes are <u>similar</u> so the angles <u>match up</u>. The angle at B corresponds with the angle at F, so they both must be 111°.

x = 111

b) Calculate the length of side AB.

The scale factor is 5 so each side in EFGH must be <u>5 times as big</u> as the corresponding side in ABCD. EF corresponds with AB.

Length of AB = Length of EF ÷ 5
= 10 cm ÷ 5 = 2 cm

Gabriel and Michael — paired up angles...

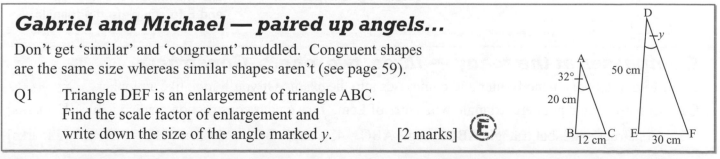

Don't get 'similar' and 'congruent' muddled. Congruent shapes are the same size whereas similar shapes aren't (see page 59).

Q1 Triangle DEF is an enlargement of triangle ABC. Find the scale factor of enlargement and write down the size of the angle marked y. [2 marks] Ⓔ

Triangle Construction

How you construct a triangle depends on what <u>info you're given</u> about the triangle...

Three sides — Use a Ruler and Compasses

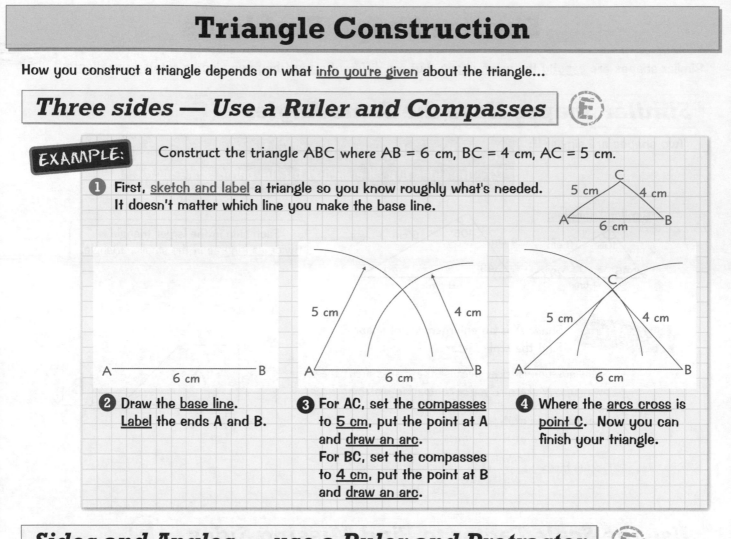

EXAMPLE: Construct the triangle ABC where AB = 6 cm, BC = 4 cm, AC = 5 cm.

❶ First, <u>sketch and label</u> a triangle so you know roughly what's needed. It doesn't matter which line you make the base line.

❷ Draw the <u>base line</u>. <u>Label</u> the ends A and B.

❸ For AC, set the <u>compasses</u> to <u>5 cm</u>, put the point at A and <u>draw an arc</u>.
For BC, set the compasses to <u>4 cm</u>, put the point at B and <u>draw an arc</u>.

❹ Where the <u>arcs cross</u> is <u>point C</u>. Now you can finish your triangle.

Sides and Angles — use a Ruler and Protractor

EXAMPLE: Construct triangle DEF. DE = 5 cm, DF = 3 cm, and angle EDF = 40°.

❶ <u>Roughly sketch and label</u> the triangle.

❷ Draw the <u>base line</u>.

❸ Draw <u>angle EDF</u> (the angle at D) — place the centre of the protractor over D, measure <u>40°</u> and put a dot.

❹ Measure <u>3 cm</u> towards the dot and label it F. Join up <u>D and F</u>. Now you've drawn the <u>two sides</u> and the <u>angle</u>. Just join up F and E to <u>complete</u> the triangle.

Compasses at the ready — three, two, one... Construct...

Don't forget to take a pencil, ruler and compasses into the exam. Or you'll look like a plonker.

Q1   Construct an equilateral triangle with sides of 5 cm. Leave your construction marks visible. [2 marks]

Q2   Construct and label triangle ABC. Angle ABC = 45°, angle BCA = 40°, side BC = 7.5 cm. [2 marks]

Loci and Constructions

A <u>LOCUS</u> (another ridiculous maths word) is simply:

A LINE or REGION that shows <u>all the points which fit in with a given rule</u>.

Make sure you learn how to do these <u>PROPERLY</u> using a <u>ruler</u> and <u>compasses</u> as shown on the next few pages.

The Four Different Types of Loci

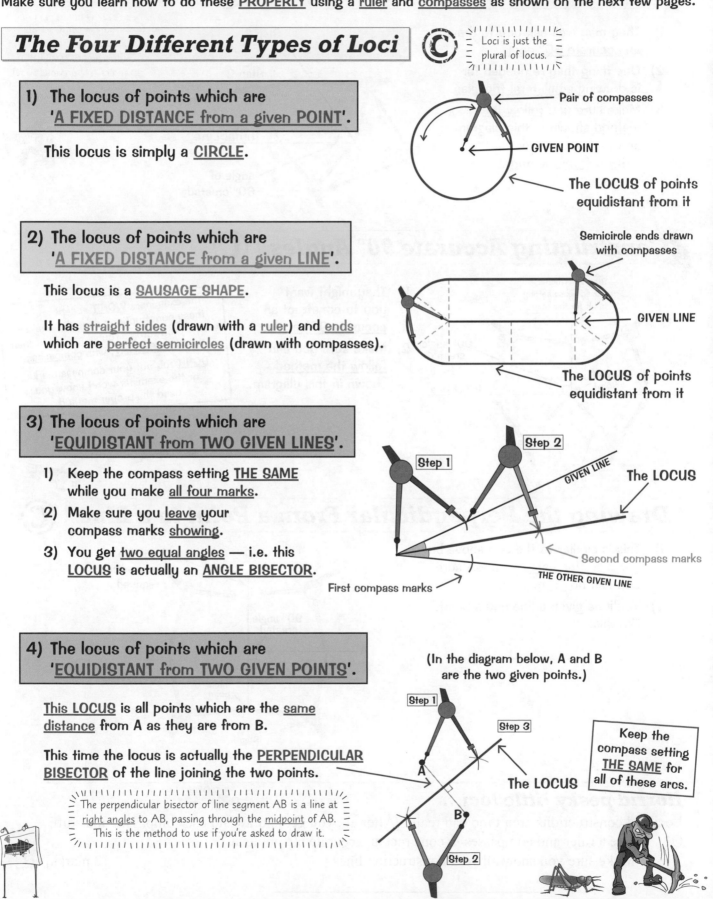

Loci is just the plural of locus.

1) The locus of points which are '<u>A FIXED DISTANCE</u> from a given <u>POINT</u>'.

This locus is simply a <u>CIRCLE</u>.

Pair of compasses

GIVEN POINT

The LOCUS of points equidistant from it

2) The locus of points which are '<u>A FIXED DISTANCE</u> from a given <u>LINE</u>'.

This locus is a <u>SAUSAGE SHAPE</u>.

It has <u>straight sides</u> (drawn with a <u>ruler</u>) and <u>ends</u> which are <u>perfect semicircles</u> (drawn with compasses).

Semicircle ends drawn with compasses

GIVEN LINE

The LOCUS of points equidistant from it

3) The locus of points which are '<u>EQUIDISTANT</u> from <u>TWO GIVEN LINES</u>'.

1) Keep the compass setting <u>THE SAME</u> while you make <u>all four marks</u>.

2) Make sure you <u>leave</u> your compass marks <u>showing</u>.

3) You get <u>two equal angles</u> — i.e. this <u>LOCUS</u> is actually an <u>ANGLE BISECTOR</u>.

Step 1   Step 2

GIVEN LINE

The LOCUS

Second compass marks

THE OTHER GIVEN LINE

First compass marks

4) The locus of points which are '<u>EQUIDISTANT</u> from <u>TWO GIVEN POINTS</u>'.

<u>This LOCUS</u> is all points which are the <u>same distance</u> from A as they are from B.

This time the locus is actually the <u>PERPENDICULAR BISECTOR</u> of the line joining the two points.

The perpendicular bisector of line segment AB is a line at <u>right angles</u> to AB, passing through the <u>midpoint</u> of AB. This is the method to use if you're asked to draw it.

(In the diagram below, A and B are the two given points.)

Step 1   Step 3

A

The LOCUS

B

Step 2

Keep the compass setting <u>THE SAME</u> for all of these arcs.

Section Five — Angles and Geometry

Loci and Constructions

Don't just read the page through once and hope you'll remember it — get your ruler, compasses and pencil out and have a go. It's the only way of testing whether you really know this stuff.

Constructing Accurate 60° Angles Ⓒ

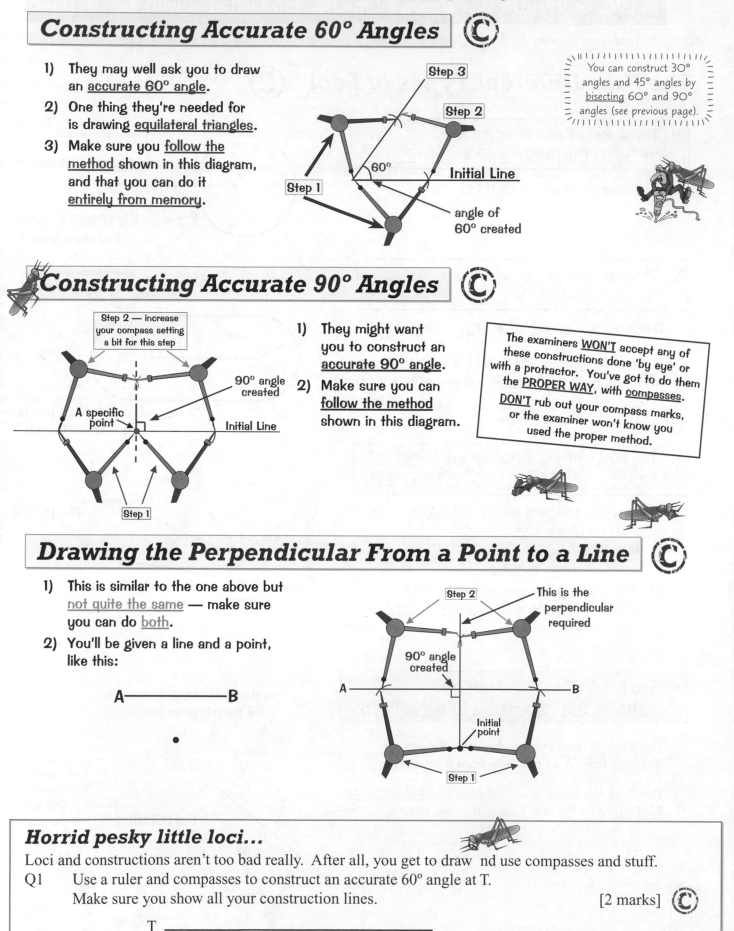

1) They may well ask you to draw an <u>accurate 60° angle</u>.

2) One thing they're needed for is drawing <u>equilateral triangles</u>.

3) Make sure you <u>follow the method</u> shown in this diagram, and that you can do it <u>entirely from memory</u>.

Step 3
Step 2
Step 1
60°
Initial Line
angle of 60° created

> You can construct 30° angles and 45° angles by <u>bisecting</u> 60° and 90° angles (see previous page).

Constructing Accurate 90° Angles Ⓒ

Step 2 — increase your compass setting a bit for this step

90° angle created

A specific point

Initial Line

Step 1

1) They might want you to construct an <u>accurate 90° angle</u>.

2) Make sure you can <u>follow the method</u> shown in this diagram.

> The examiners <u>WON'T</u> accept any of these constructions done 'by eye' or with a protractor. You've got to do them the <u>PROPER WAY</u>, with <u>compasses</u>.
> <u>DON'T</u> rub out your compass marks, or the examiner won't know you used the proper method.

Drawing the Perpendicular From a Point to a Line Ⓒ

1) This is similar to the one above but <u>not quite the same</u> — make sure you can do <u>both</u>.

2) You'll be given a line and a point, like this:

A———————B

•

Step 2
This is the perpendicular required
90° angle created
A ————————————— B
Initial point
Step 1

Horrid pesky little loci...

Loci and constructions aren't too bad really. After all, you get to draw nd use compasses and stuff.

Q1 Use a ruler and compasses to construct an accurate 60° angle at T.
Make sure you show all your construction lines. **[2 marks]** Ⓒ

T ———————————————————

Loci and Constructions — Worked Example

Now you know what <u>loci</u> are, and how to do all the <u>constructions</u> you need, it's time to put them all together.

Finding a Locus That Satisfies More Than One Rule Ⓒ

In the exam, you might be given <u>two conditions</u>, and asked to find the region that satisfies <u>both</u> of them.

EXAMPLE:

Mary is deciding where to plant a tree in her garden.
She makes a plan of her garden with a scale of <u>1 cm = 2 m</u>.

1) Her house runs along <u>side AD</u>. The tree cannot be planted <u>within 4 m</u> of the house.
2) Mary wants the tree to be <u>closer to corner D than to corner B</u>.

Complete the plan to show where Mary can plant the tree.

There's more on scale drawings on pages 97-98.

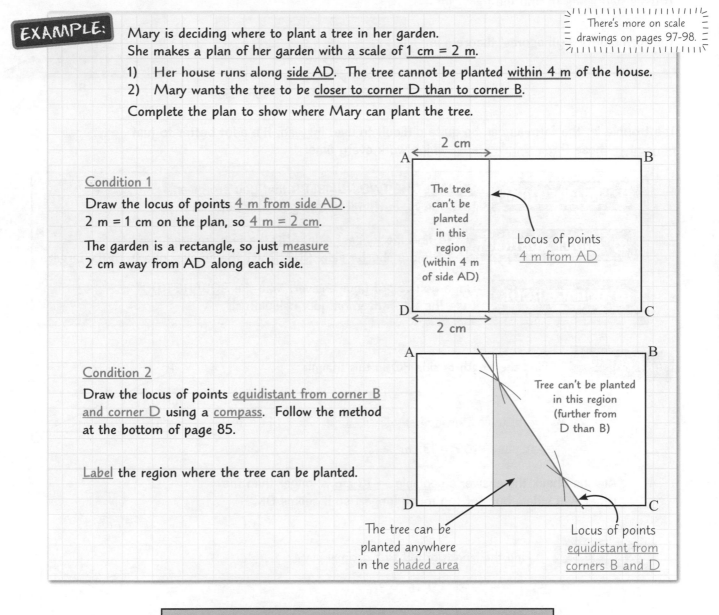

<u>Condition 1</u>

Draw the locus of points <u>4 m from side AD</u>.
2 m = 1 cm on the plan, so <u>4 m = 2 cm</u>.

The garden is a rectangle, so just <u>measure</u>
2 cm away from AD along each side.

<u>Condition 2</u>

Draw the locus of points <u>equidistant from corner B and corner D</u> using a <u>compass</u>. Follow the method at the bottom of page 85.

<u>Label</u> the region where the tree can be planted.

Always <u>leave your construction lines showing</u>.
They show the examiner that you used the proper method.

Stay at least 3 m away from point C — or I'll release the hounds...

I can't stress this enough — make sure you draw your diagrams ACCURATELY
(using a ruler and a pair of compasses). Now try this Exam Practice Question:

Q1 A room in a stately home is shown in the diagram.
Visitors must stay at least 2 m away from the portrait and at least 2 m
away from the statue. Make a copy of the diagram using a scale of
1 cm = 1 m and indicate on it the area where visitors can go.

[4 marks]

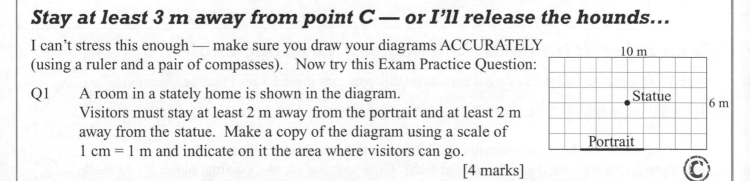

Pythagoras' Theorem

Once upon a time there lived a clever chap called Pythagoras. He came up with a clever theorem...

Pythagoras' Theorem is Used on Right-Angled Triangles Ⓒ

Pythagoras' theorem only works for <u>RIGHT-ANGLED TRIANGLES</u>.
It uses <u>two sides</u> to find the <u>third side</u>.

The formula for Pythagoras' theorem is:

$$a^2 + b^2 = c^2$$

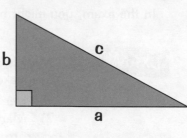

short sides   long side

The trouble is, the formula can be quite difficult to use. <u>Instead</u>, it's a lot better to just <u>remember</u> these <u>three simple steps</u>, which work every time:

1) SQUARE THEM — SQUARE THE TWO NUMBERS that you are given, (use the x^2 button if you've got your calculator.)

2) ADD or SUBTRACT — To find the <u>longest side</u>, <u>ADD</u> the two squared numbers. To find <u>a shorter side</u>, <u>SUBTRACT</u> the smaller one from the larger.

3) SQUARE ROOT — Once you've got your answer, take the <u>SQUARE ROOT</u> (use the $\sqrt{}$ button on your calculator).

EXAMPLE: Find the length of side PQ in this triangle.

1) <u>Square</u> them: $5^2 = 25$, $12^2 = 144$

2) You want to find the <u>longest side</u>, so <u>ADD</u>: $25 + 144 = 169$

3) <u>Square root</u>: $\sqrt{169} = 13$ cm ←

Always check the answer's <u>sensible</u> — <u>13 cm</u> is longer than the other two sides, but not too much longer, so it seems OK.

EXAMPLE: Find the length of SU to 1 decimal place.

1) <u>Square</u> them: $3^2 = 9$, $6^2 = 36$

2) You want to find <u>a shorter side</u>, so <u>SUBTRACT</u>: $36 - 9 = 27$

3) <u>Square root</u>: $\sqrt{27} = 5.196...$
                $= 5.2$ m (to 1 d.p.)

Check the answer is <u>sensible</u> — yes, it's a bit shorter than the longest side.

Remember, if it's not a right angle, it's a wrong angle...

Once you've learned all the Pythagoras facts on this page, try these Exam Practice Questions.

Q1   Find the length of AC correct to 1 decimal place.   [3 marks] Ⓒ

Q2   A 4 m long ladder leans against a wall. Its base is 1.2 m from the wall.
      How far up the wall does the ladder reach? Give your answer to 1 decimal place.   [3 marks] Ⓒ

Revision Questions for Section Five

There are lots of opportunities to show off your artistic skills here (as long as you use them to answer the questions).

- Try these questions and <u>tick off each one</u> when you <u>get it right</u>.
- When you've done <u>all the questions</u> for a topic and are <u>completely happy</u> with it, tick off the topic.

Angles and Polygons (p71-78)

1) What is the name for an angle larger than 90° but smaller than 180°?

2) What do angles in a quadrilateral add up to?

3) Find the missing angles in the diagrams below.

a) 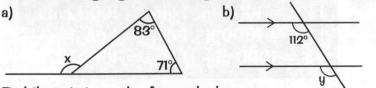 b) 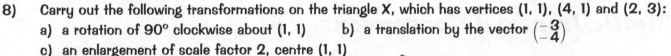 c)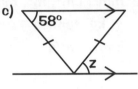

4) Find the exterior angle of a regular hexagon.

5) Find the sum of interior angles in a regular 12-sided polygon.

6) Why do hexagons tessellate but pentagons don't?

Transformations and Similar Shapes (p79-83)

7) Describe the transformation that maps:
   a) Shape A onto Shape B
   b) Shape A onto shape C

8) Carry out the following transformations on the triangle X, which has vertices (1, 1), (4, 1) and (2, 3):
   a) a rotation of 90° clockwise about (1, 1)   b) a translation by the vector $\binom{-3}{-4}$
   c) an enlargement of scale factor 2, centre (1, 1)

9) These two triangles are similar.
   Write down the values of b and y.

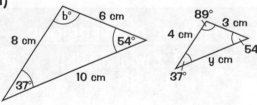

Constructions and Loci (p84-87)

10) Construct triangle XYZ, where XY = 5.6 cm, XZ = 7.2 cm and angle YXZ = 55°.

11) What shape does the locus of points that are a fixed distance from a given point make?

12) Construct an accurate 45° angle.

13) Draw a square with sides of length 6 cm and label it ABCD. Shade the region
    that is nearer to AB than CD and less than 4 cm from vertex A.

Pythagoras' Theorem (p88)

14) What is the formula for Pythagoras' theorem? What type of triangle can you use it on?

15) A museum has a flight of stairs up to its front door (see diagram).
    A ramp is to be put over the top of the steps for wheelchair users.
    Calculate the length that the ramp would need to be to 1 d.p.

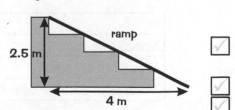

16) A rectangle has a diagonal of 15 cm. Its short side is 4 cm.
    Calculate the length of the rectangle's long side to 1 d.p.

17) Find the distance between the points A(2, 5) and B(5, 1).

Section Five — Angles and Geometry

Metric and Imperial Units

OK, I admit it, this page is packed full of facts and figures, but it's all important stuff. You need to know what the different types of units are used for, and the rules for converting between them.

Metric Units (F)

① Length   mm, cm, m, km
② Area   mm², cm², m², km²,
③ Volume   mm³, cm³, m³, ml, litres
④ Weight   g, kg, tonnes
⑤ Speed   km/h, m/s

MEMORISE THESE KEY FACTS:

1 cm = 10 mm	1 tonne = 1000 kg
1 m = 100 cm	1 litre = 1000 ml
1 km = 1000 m	1 litre = 1000 cm³
1 kg = 1000 g	1 cm³ = 1 ml
1 g = 1000 mg	1 litre = 100 cl

You need to know these off by heart for the exam.

Imperial Units (F)

① Length   Inches, feet, yards, miles
② Area   Square inches, square feet, square miles
③ Volume   Cubic inches, cubic feet, pints, gallons
④ Weight   Ounces, pounds, stones, tons
⑤ Speed   mph

IMPERIAL UNIT CONVERSIONS
1 Foot = 12 Inches
1 Yard = 3 Feet
1 Gallon = 8 Pints
1 Stone = 14 Pounds (lb)
1 Pound = 16 Ounces (oz)

You don't need to know these for the exam, but you should be able to use the conversions.

Metric-Imperial Conversions (F)

APPROXIMATE CONVERSIONS
1 kg ≈ 2.2 pounds (lb)
1 foot ≈ 30 cm
1 litre ≈ 1¾ pints
1 gallon ≈ 4.5 litres
1 mile ≈ 1.6 km (or 5 miles ≈ 8 km)

≈ means roughly equal to

YOU NEED TO LEARN THESE —
they don't promise to give you these
in the exam and if they're feeling
mean (as they often are), they won't.

Imperial units — they're mint...

Learn the metric and metric-imperial conversions, then <u>cover</u> the page and try to <u>scribble</u> them all down. You'll get your chance to put them into practice on the next page, but for now have a go at this question...

Q1   Fill in the gaps in this table by writing a sensible unit for each measurement.

	Metric	Imperial
Speed of a train	km/h	
Weight of an apple		ounces
Volume of a bottle of milk	litres	

[3 marks] (F)

Converting Units

Conversion factors are a really good way of dealing with all sorts of questions — and the method is dead easy. Prepare yourself for the <u>3 steps</u> towards conversion-related happiness...

3 Step Method (E)

① Find the <u>conversion factor</u> (always easy).

② <u>Multiply AND divide by it</u>.

③ Choose the <u>common sense answer</u>.

Examples (E)

1 A giant sea slug called Kevin was washed up near Grange-over-Sands. He was 18.6 m in length. How long is this in cm?

1) Find the <u>conversion factor</u>

1 m = 100 cm

Conversion factor = 100

2) <u>Multiply and divide</u> by it

18.6 × 100 = 1860 — makes sense
18.6 ÷ 100 = 0.186 — ridiculous

3) Choose the <u>sensible answer</u>

18.6 m = **1860 cm**

2 The villages of Chickenham and Frogville are 12 miles from each other. How far is this in km?

1 mile ≈ 1.6 km

1) Find the <u>conversion factor</u> ⟶ Conversion factor = 1.6

2) <u>Multiply and divide</u> by it — 12 × 1.6 = 19.2

3) Choose the <u>sensible answer</u> — 12 ÷ 1.6 = 7.5
1 mile is about 1.6 km so there should be <u>more km than miles</u>

12 miles ≈ **19.2 km**

3 Lisa buys a carton of her favourite gooseberry and lime juice. The carton has a volume of 1500 cm^3. What is its volume in pints?

1) First <u>convert cm^3 to litres</u> — the numbers are simple here.

$1000 \text{ cm}^3 = 1 \text{ litre} \longrightarrow 1500 \text{ cm}^3 = 1.5 \text{ litres}$

2) Then <u>convert litres to pints</u> — use the <u>conversion factor</u>.

1 litre ≈ 1.75 pints ⟶ Conversion factor = 1.75

3) <u>Multiply and divide</u> by it.

1.5 × 1.75 = 2.625
1.5 ÷ 1.75 = 0.8571...

4) Choose the <u>sensible answer</u> — 1 litre is about 1.75 pints so there should be <u>more pints than litres</u>.

1.5 litres ≈ **2.625 pints**

There are 3 steps to conversion heaven...

Once you've got the 3-step method well and truly sorted, have a go at these Exam Practice Questions.

Q1 Jamie buys a pack of three tennis balls. They weigh 60 g each. What is their total weight in kg? **[3 marks]** (E)

Q2 Lisa pours 263 ml of gooseberry and lime juice from the 1.5 litre carton. How many ml are left in the carton? **[2 marks]** (E)

More Conversions

This page is a bit <u>trickier</u>, but everything comes down to the same old <u>conversion factors</u>...

Converting Speeds

Don't panic if you're asked to convert a <u>speed</u> from, say, miles per hour (mph) to km per hour (km/h)...

...if the <u>time part</u> of the units <u>stays the same</u> then it's really just a <u>distance conversion</u> in disguise.

EXAMPLE:

Sophie is driving at 60 km/h. The speed limit is 40 mph. Is she breaking the speed limit?

① First <u>convert</u> 60 km into miles:

1 mile $\approx$ 1.6 km ⟶ Conversion factor = 1.6

$60 \times 1.6 = 96$ — too big

$60 \div 1.6 = 37.5$ — makes sense, so 60 km = 37.5 miles

② Add in the '<u>per hour</u>' bit to get the <u>speed</u>:

60 km/h = 37.5 mph

So, Sophie **isn't breaking the 40 mph speed limit.**

Converting Areas and Volumes — Tricky (D)

Be really <u>careful</u> — 1 m = 100 cm <u>**DOES NOT**</u> mean 1 m² = 100 cm² or 1 m³ = 100 cm³.

You won't slip up if you <u>**LEARN THESE RULES**</u>:

<u>Area</u>: units come with a <u>**2**</u>, e.g. mm², cm², m² — <u>use the conversion factor 2 times</u>.	<u>Volume</u>: units come with a <u>**3**</u>, e.g. mm³, cm³, m³ — <u>use the conversion factor 3 times</u>.

<u>Multiply AND divide</u> the correct number of times, then pick the <u>sensible</u> answer — use the <u>rule</u> linking the units to decide whether your answer should be <u>bigger or smaller</u> than what you started with.

EXAMPLES:

1. The area of the top of a table is 0.6 m². Find its area in cm².

1) Find the <u>conversion factor</u>:

1 m = 100 cm ⟶ Conversion factor = 100

2) It's an area — multiply and divide <u>twice</u> by conversion factor:

$0.6 \times 100 \times 100 = 6000$ — makes sense

$0.6 \div 100 \div 100 = 0.00006$ — too small

3) Choose the <u>sensible</u> answer:

0.6 m² = **6000 cm²**

1 m = 100 cm so expect more cm than m.

2. A glass has a volume of 72 000 mm³. What is its volume in cm³?

1) Find the <u>conversion factor</u>:

1 cm = 10 mm ⟶ Conversion factor = 10

2) It's a volume — multiply and divide <u>3 times</u> by conversion factor:

$72\,000 \times 10 \times 10 \times 10 = 72\,000\,000$ — too big

$72\,000 \div 10 \div 10 \div 10 = 72$ — makes sense

3) Choose the <u>sensible</u> answer:

72 000 mm³ = **72 cm³**

1 cm = 10 mm, so expect fewer cm than mm.

You don't have the Conversion Factor. I thought it was very karaoke...

Make sure you've got to grips with the examples on this page, then have a go at these practice questions.

Q1 A pizza has an area of 22 530 mm². Find its area in cm². [2 marks] (D)

Q2 Tom rides his bicycle at an average speed of 10 km/h. What speed is this in mph? [2 marks] (D)

Reading Scales

You can pick up some <u>easy marks</u> if you get a question asking you to read a scale. The <u>same rules</u> apply to scales measuring <u>lengths</u>, <u>weights</u>, <u>volumes</u>, <u>speeds</u> and <u>temperatures</u>, so learn them and those marks are yours.

How to Read a Scale Ⓖ

All scales consist of a <u>line divided into intervals</u> like this:

The line this arrow's pointing to is 3 small gaps after 30.

0 10 20 30 40 cm

Large gap between numbers

Small gap

units (these won't always be units of length)

To <u>read a point</u> on the scale (e.g. where the orange arrow's pointing to), you need to know <u>what each small gap represents</u>:

$$\text{Small gap} = \frac{\text{Size of large gap between numbers}}{\text{Number of small gaps between numbers}}$$

Make sure you count the gaps — <u>DON'T</u> count the small marks or you'll get the wrong answer.

1) On the scale above there's a <u>difference of 10</u> between the numbers, and <u>5 small gaps</u> between them, so each small gap's worth 10 ÷ 5 = 2 cm.

2) The <u>orange arrow's</u> 3 small gaps after 30. 3 small gaps = 3 × 2 = 6, so it's pointing to 30 + 6 = 36 cm.

EXAMPLES:

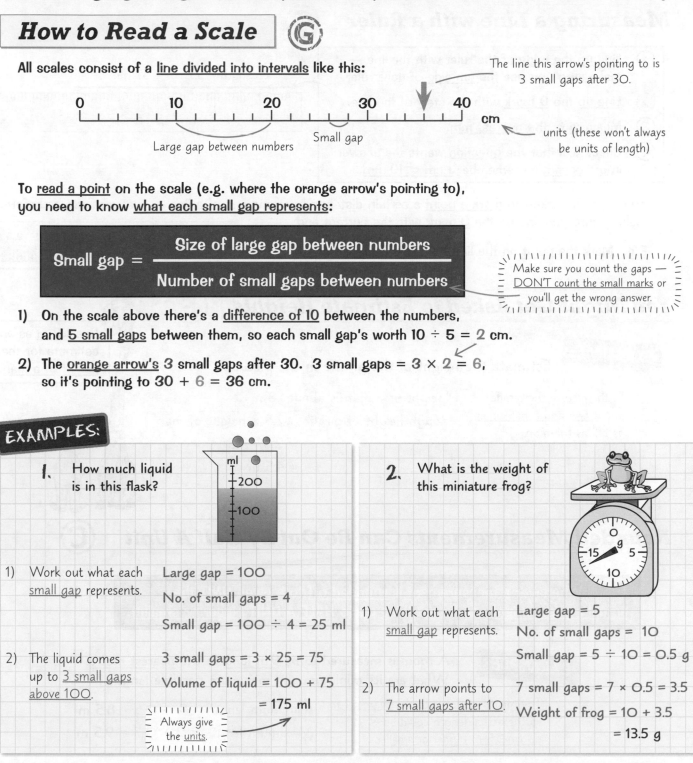

1. How much liquid is in this flask?

ml
200
100

1) Work out what each <u>small gap</u> represents.

Large gap = 100

No. of small gaps = 4

Small gap = 100 ÷ 4 = 25 ml

2) The liquid comes up to <u>3 small gaps above 100</u>.

3 small gaps = 3 × 25 = 75

Volume of liquid = 100 + 75

= 175 ml

Always give the <u>units</u>.

2. What is the weight of this miniature frog?

g
15   5
10

1) Work out what each <u>small gap</u> represents.

Large gap = 5

No. of small gaps = 10

Small gap = 5 ÷ 10 = 0.5 g

2) The arrow points to <u>7 small gaps after 10</u>.

7 small gaps = 7 × 0.5 = 3.5

Weight of frog = 10 + 3.5

= 13.5 g

There's nothing like curling up on the sofa and reading a good scale...

You need to be comfortable reading any sort of scale, so give these Exam Practice Questions a try.

Q1   What speed does the speedometer show?

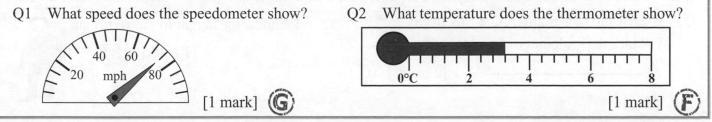

20   mph   80
40   60

[1 mark] Ⓖ

Q2   What temperature does the thermometer show?

0°C   2   4   6   8

[1 mark] Ⓕ

Rounding and Estimating Measurements

You need to be able to <u>measure things accurately</u>, and <u>make estimates</u> of lengths and heights. There's nothing too tricky here, so have a good read through and you'll be sorted.

Measuring a Line with a Ruler Ⓖ

① <u>Line up</u> the <u>edge</u> of the ruler with the line — make sure you use the <u>cm side</u> of your ruler.

② <u>Line up</u> the <u>0 mark</u> with the <u>start</u> of the line.

③ <u>Measure</u> to the <u>nearest mm</u>.

④ Check whether the <u>question</u> wants the answer in <u>cm or mm</u> — remember <u>1 cm = 10 mm</u>.

Length = 5.7 cm = 57 mm

0 cm 1 2 3 4 5 6 7

You might be asked to <u>mark a point</u> a certain distance from one end of a line — just measure as before, making sure you line up the <u>0 mark</u> with the <u>correct end</u>.

E.g. Mark the point on the line XY that's 3 cm from point Y.

0 cm 1 2 3 4 5 6

X ——×—— Y

You Might Get Asked to Estimate Heights Ⓕ

Use <u>1.8 m</u> as an estimate for the <u>height of a man</u>.

EXAMPLE: Estimate the height of the giraffe in the picture.

In the picture the giraffe's about <u>two and a half times</u> as tall as the man.

Height of a man is about 1.8 m
Rough height of giraffe = 2.5 × height of man
= 2.5 × 1.8
= 4.5 m

Rounded Measurements Can Be Out By Half A Unit Ⓒ

Whenever a measurement is <u>rounded off</u> to a <u>given UNIT</u> the <u>actual measurement</u> can be anything up to <u>HALF A UNIT bigger or smaller</u>.

EXAMPLE: A room is measured to be <u>9 m long</u> to the <u>nearest metre</u>. What are its minimum and maximum possible lengths?

The measurement is to the <u>nearest 1 m</u>, so the actual length could be <u>up to 0.5 m bigger or smaller</u>.

Minimum length = 9 − 0.5 = 8.5 m
Maximum length = 9 + 0.5 = 9.5 m

Don't underestimate the importance of eating biscuits when revising...

Once you've been through everything on this page, have a go at these questions to make sure you're all set.

Q1 a) Measure the length of line MN in cm.
b) Mark the midpoint of line MN with a cross.

M ——————————— N   [2 marks] Ⓖ

Q2   Estimate the height of the snowman.

[1 mark] Ⓕ

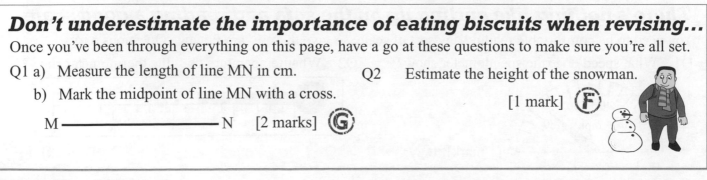

Reading Timetables

I'm sure you're a dab hand at reading <u>clocks</u>, but here's a quick reminder...

<u>am</u> means <u>morning</u>
<u>pm</u> means <u>afternoon or evening</u>.

<u>12 am</u> (OO:OO) means <u>midnight</u>.
<u>12 pm</u> (12:OO) means <u>noon</u>.

12-hour clock	24-hour clock
12.OO am	OO:OO
1.12 am	O1:12
12.15 pm	12:15
1.47 pm	13:47
11.32 pm	23:32

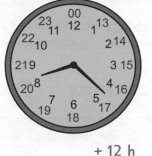

The hour parts of times on 12- and 24- hour clocks are <u>different after 1 pm</u>:
<u>add 12 hours</u> to go from <u>12-hour to 24-hour</u>, and subtract 12 to go the other way.

3.24 pm $\xrightarrow{+ 12 \text{ h}}$ 15:24
$\xleftarrow{- 12 \text{ h}}$

Break Time Calculations into Simple Stages (F)

EXAMPLE: Angela watched a film that started at 7.2O pm and finished at 1O.O5 pm. How long was the film in minutes?

1) Split the time between 7.2O pm and 1O.O5 pm into <u>simple stages</u>.

7.2O pm $\xrightarrow{+ 2 \text{ hours}}$ 9.2O pm $\xrightarrow{+ 4O \text{ minutes}}$ 1O.OO pm $\xrightarrow{+ 5 \text{ minutes}}$ 1O.O5 pm

2) <u>Convert</u> the hours to minutes.   2 hours = 2 × 6O = 12O minutes

3) <u>Add</u> to get the total minutes.   12O + 4O + 5 = **165 minutes**

> <u>Avoid calculators</u> — the decimal answers they give are confusing, e.g. <u>2.5 hours = 2 hours 3O mins</u>, <u>NOT 2 hours 5O mins</u>.

Timetable Exam Questions (E)

EXAMPLE:

Use the timetable to answer these questions:

a) How long does it take for the bus to get from <u>Market Street</u> to the hospital?

Bus Timetable				
Bus Station	18 45	19 OO	19 15	19 3O
Market Street	18 52	19 O7	19 22	19 37
Long Lane Shops	19 O1	19 16	19 31	19 46
Train Station	19 11	19 26	19 41	19 56
Hospital	19 23	19 38	19 53	2O O8

Read times from the <u>same column</u> (I've used the 1st) — break the <u>time</u> into <u>stages</u>.

Market Street   Hospital
18:52 $\xrightarrow{+ 8 \text{ mins}}$ 19:OO $\xrightarrow{+ 23 \text{ mins}}$ 19:23   8 + 23 = **31 minutes**

b) Harry wants to get a bus from the <u>bus station</u> to the <u>train station</u> in time for a train that leaves at <u>19:3O</u>. What is the latest bus he can catch?

1) Read along the <u>train station</u> row.

2) Move up this column to the <u>bus station</u> row and read off the entry.

19 11   (19 26)   19 41   19 56

This is the latest time he could arrive before 19:3O.

The bus that gets to the train station at 19:26 leaves the bus station at **19:OO**.

BREAKING NEWS: Public panic after warning over calculator use...

Have a go at these Exam Practice Questions and use the timetable above to answer Q2.

Q1 A plane takes off at 9.37 am and lands at 11.16 am. How long is the flight in minutes? [1 mark] (F)

Q2 Amy lives 1O minutes' walk from the Market Street bus stop. She wants to be at Long Lane shops by 19:45. Write a schedule for her journey from her house to the shops. [2 marks] (E)

Compass Directions and Bearings

Compass points and bearings both describe the <u>direction</u> of something.
You'll have seen a <u>compass</u> before — make sure you <u>know</u> all <u>8 directions</u>.

<u>Bearings</u> are trickier but really useful — they can
describe <u>any direction</u>, not just the 8 compass points.

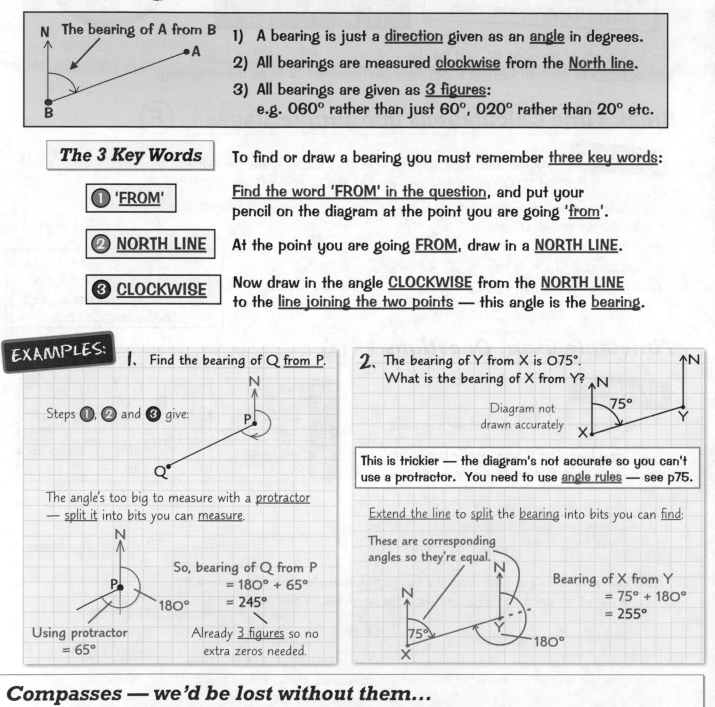

Bearings (D)

1) A bearing is just a <u>direction</u> given as an <u>angle</u> in degrees.

2) All bearings are measured <u>clockwise</u> from the <u>North line</u>.

3) All bearings are given as <u>3 figures</u>:
   e.g. 060° rather than just 60°, 020° rather than 20° etc.

The 3 Key Words

To find or draw a bearing you must remember <u>three key words</u>:

1 'FROM' — <u>Find the word 'FROM' in the question</u>, and put your pencil on the diagram at the point you are going '<u>from</u>'.

2 NORTH LINE — At the point you are going <u>FROM</u>, draw in a <u>NORTH LINE</u>.

3 CLOCKWISE — Now draw in the angle <u>CLOCKWISE</u> from the <u>NORTH LINE</u> to the <u>line joining the two points</u> — this angle is the <u>bearing</u>.

EXAMPLES:

1. Find the bearing of Q <u>from</u> P.

Steps **1**, **2** and **3** give:

The angle's too big to measure with a <u>protractor</u>
— <u>split it</u> into bits you can <u>measure</u>.

Using protractor
= 65°

So, bearing of Q from P
= 180° + 65°
= 245°

Already <u>3 figures</u> so no extra zeros needed.

2. The bearing of Y from X is 075°.
What is the bearing of X from Y?

Diagram not drawn accurately

This is trickier — the diagram's not accurate so you can't use a protractor. You need to use <u>angle rules</u> — see p75.

<u>Extend the line</u> to <u>split</u> the <u>bearing</u> into bits you can <u>find</u>:

These are corresponding angles so they're equal.

Bearing of X from Y
= 75° + 180°
= 255°

Compasses — we'd be lost without them...

Here are some bearings questions to try. You might find the North line's drawn for you in the exam.

Q1 Measure the bearing of C from D.

[2 marks] (D)

Q2 What is the bearing of H from G?

[2 marks] (D)

Diagram not drawn accurately

Maps and Scale Drawings

<u>Scales</u> tell you what a <u>distance</u> on a <u>map</u> or <u>drawing</u> represents in <u>real life</u>. They can be written in various ways, but they all boil down to something like "<u>1 cm represents 5 km</u>".

Map Scales

1 cm = 3 km — "1 cm represents 3 km"

1 : 2000 — 1 cm on the map means 2000 cm in real life.
Converting to m gives "1 cm represents 20 m".

```
|_____|
0   km   1
```
Use a ruler — the line's 2 cm long, so 2 cm means 1 km.
Dividing by 2 gives "1 cm represents 0.5 km".

See p91 for a reminder about conversions.

To <u>convert</u> between <u>maps</u> and <u>real life</u>, <u>learn</u> these rules:

- To find <u>REAL-LIFE</u> distances, <u>MULTIPLY</u> by the <u>MAP SCALE</u>.
- To find <u>MAP</u> distances, <u>DIVIDE</u> by the <u>MAP SCALE</u>.
- Make sure your map scale is of the form "<u>1 cm = ...</u>"
- Always check your answer looks <u>sensible</u>.

Converting from Map Distance to Real Life — Multiply Ⓔ

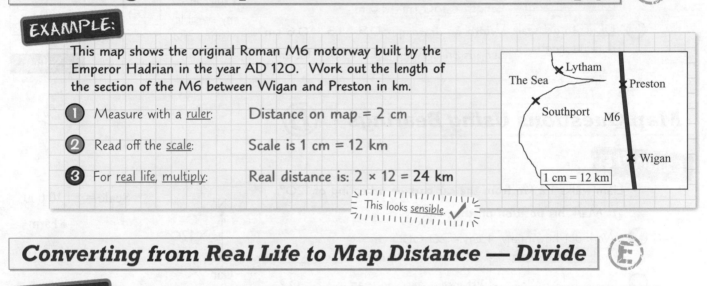

EXAMPLE:

This map shows the original Roman M6 motorway built by the Emperor Hadrian in the year AD 120. Work out the length of the section of the M6 between Wigan and Preston in km.

① Measure with a <u>ruler</u>: Distance on map = 2 cm

② Read off the <u>scale</u>: Scale is 1 cm = 12 km

③ For <u>real life</u>, multiply: Real distance is: 2 × 12 = 24 km

This looks <u>sensible</u>. ✓

Converting from Real Life to Map Distance — Divide Ⓔ

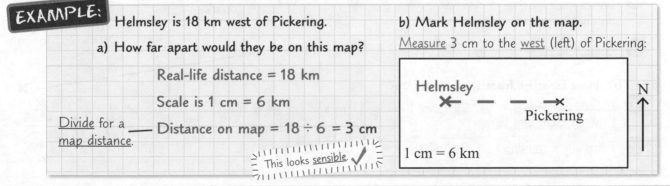

EXAMPLE:

Helmsley is 18 km west of Pickering.

a) How far apart would they be on this map?

Real-life distance = 18 km

Scale is 1 cm = 6 km

<u>Divide</u> for a <u>map distance</u>. — Distance on map = 18 ÷ 6 = 3 cm

This looks <u>sensible</u>. ✓

b) Mark Helmsley on the map.

<u>Measure</u> 3 cm to the <u>west</u> (left) of Pickering:

Helmsley
×– – – – –×
       Pickering

1 cm = 6 km   N ↑

Follow this map of the road to exam glory... What?... Cheesy?... Me?

Give these practice questions a go once you're happy with the two formulas and know when to use them.

Q1 Use this map to find the distance between Broughton and Coniston in miles.
[2 marks] Ⓔ

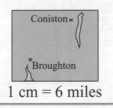

Coniston ×

× Broughton

1 cm = 6 miles

Q2 Sarah's house is 2.25 km away from Luke's house. How far apart in cm would they be on a map where 1 cm represents 500 m?
[2 marks] Ⓔ

Maps and Scale Drawings

Scale Drawings (E)

Scale drawings work just like maps. To convert between real life and scale drawings, just replace the word 'map' with 'drawing' in the rules on the previous page.

EXAMPLE:

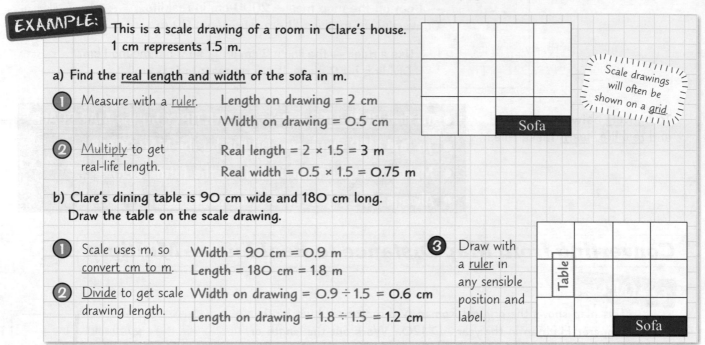

This is a scale drawing of a room in Clare's house.
1 cm represents 1.5 m.

a) Find the real length and width of the sofa in m.

① Measure with a ruler.   Length on drawing = 2 cm
                          Width on drawing = 0.5 cm

② Multiply to get   Real length = 2 × 1.5 = 3 m
   real-life length.   Real width = 0.5 × 1.5 = 0.75 m

Scale drawings will often be shown on a grid.

b) Clare's dining table is 90 cm wide and 180 cm long.
   Draw the table on the scale drawing.

① Scale uses m, so   Width = 90 cm = 0.9 m
   convert cm to m.   Length = 180 cm = 1.8 m

② Divide to get scale   Width on drawing = 0.9 ÷ 1.5 = 0.6 cm
   drawing length.   Length on drawing = 1.8 ÷ 1.5 = 1.2 cm

③ Draw with a ruler in any sensible position and label.

Map Questions Using Bearings (D)

EXAMPLE:

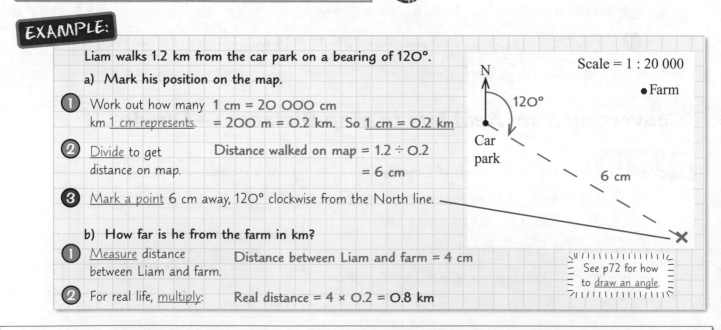

Liam walks 1.2 km from the car park on a bearing of 120°.

a) Mark his position on the map.

① Work out how many   1 cm = 20 000 cm
   km 1 cm represents.   = 200 m = 0.2 km. So 1 cm = 0.2 km

② Divide to get   Distance walked on map = 1.2 ÷ 0.2
   distance on map.   = 6 cm

③ Mark a point 6 cm away, 120° clockwise from the North line.

b) How far is he from the farm in km?

① Measure distance   Distance between Liam and farm = 4 cm
   between Liam and farm.

② For real life, multiply: Real distance = 4 × 0.2 = 0.8 km

Scale = 1 : 20 000

See p72 for how to draw an angle.

Well, you should have got your bearings on map scales by now...

Keep your ruler and protractor handy when you're doing map and scale drawing questions.

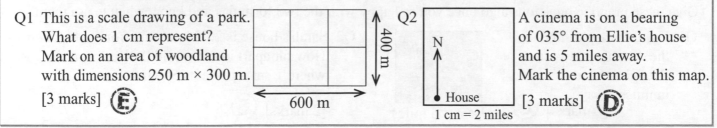

Q1 This is a scale drawing of a park.
    What does 1 cm represent?
    Mark on an area of woodland
    with dimensions 250 m × 300 m.
    [3 marks] (E)

400 m

600 m

Q2 A cinema is on a bearing
    of 035° from Ellie's house
    and is 5 miles away.
    Mark the cinema on this map.
    [3 marks] (D)

N

House

1 cm = 2 miles

Speed

Learn the formula triangle on this page and you'll be ready to tackle any speed question that comes along...

Speed = Distance ÷ Time ⓓ

This is the basic formula for calculating speed from distance and time:

$$SPEED = \frac{DISTANCE}{TIME}$$

You also need to be able to find distance from speed and time, and time from speed and distance. But fear not, there's no need for any algebra — it all becomes simple if you use the formula triangle...

Using the Formula Triangle ⓓ

Use the words SaD Times to help you remember the order of the letters (S^DT).

So if it's a question on speed, distance and time just say: SAD TIMES

To use the formula triangle, cover up the thing you want to find and write down what's left showing.

To find SPEED — cover S: To find DISTANCE — cover D: To find TIME — cover T:

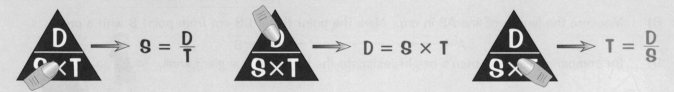

$$S = \frac{D}{T} \qquad D = S \times T \qquad T = \frac{D}{S}$$

Always give units when you do speed calculations — the units you get out depend on the units you put in, e.g. distance in km and time in hours gives speed in km per hour (km/h).

EXAMPLES:

1. Rob cycles 18 miles in 2 hours. What is his average speed?

1) Write down the formula triangle.

2) You want speed so covering S gives: $S = \frac{D}{T}$

3) Put in the numbers. $S = 18 \div 2$

4) Give the units. $= 9$ mph

Miles and hours go in so miles per hour comes out.

2. A cheetah runs at a constant speed of 27 m/s for 20 s. What distance does it cover?

1) Write down the formula triangle.

2) You want distance so covering D gives: $D = S \times T$

3) Put in the numbers. $D = 27 \times 20$

4) Give the units. $= 540$ m

m/s and s go in so m comes out.

Formula triangles — it's all a big cover-up...

Once you're happy with how these examples work, have a go at these Exam Practice Questions.

Q1   A car travels 9 miles at 36 miles per hour. How many minutes does it take?   [2 marks] ⓓ

Q2   Rachel walks 4.5 km in 1 hour 30 minutes and Tyler walks 7 km in 2 hours.
Whose average speed is the fastest? Show all your working.   [4 marks] ⓒ

Revision Questions for Section Six

When you think you've got the measure of this section, use this page to make sure it's all sorted.

- Try these questions and <u>tick off each one</u> when you <u>get it right</u>.
- When you've done <u>all the questions</u> for a topic and are <u>completely happy</u> with it, tick off the topic.

Converting Units (p90-92) ☑

1) Fill in these gaps.

1 m = ... cm	1 litre ≈ .... pints
1 litre = ... cm³	1 mile ≈ .... km
1000 m = ... km	1 foot ≈ .... cm

2) What is the 3-step method for converting units?

3) Kevin is filling in a form to join the gym and needs to give his weight in kg. He knows he weighs 143 pounds. How much is this in kg?

4) How many times do you multiply or divide by the conversion factor when converting the units of an area? How many times for a volume?

5) A bath holds 120 litres of water. What is its volume in m³?

Reading Scales and Measuring (p93-94) ☑

6) How do you work out what a small gap on a scale stands for?

7) Using the ruler shown, what is the length of this key?

8) Measure the length of line AB in cm. Mark the point that's 1.3 cm from point B with a cross.

A ——————————————— B

9) By comparing with the man's height, estimate the height of this giant snail.

Reading Timetables (p95) ☑

10) Write a) 4.20 pm as a 24-hour clock time, b) 07:52 as a 12-hour clock time.

11) Using the timetable, how many minutes does the journey from Edinburgh to York last for?

12) Jane lives in Berwick and needs to be in Durham by 1.30 pm. a) What is the latest train she can catch? She lives 20 minutes' walk from the train station. b) What is the latest time she should leave the house?

Train Timetable			
Edinburgh	11 14	11 37	12 04
Berwick	11 55	12 18	12 45
Newcastle	12 43	13 06	13 33
Durham	12 57	13 20	13 47
Darlington	13 15	13 38	14 05
York	13 45	14 08	14 35

Bearings, Maps and Scale Drawings (p96-98) ☑

13) What are the three key words to remember when you're working with bearings?

14) How do you use a map scale to go from a real-life distance to a distance on a map, and vice versa?

15) Bobby is planning the layout of a new car park for his local supermarket, shown on the right. Draw a plan of the car park using a scale of 1 cm = 5 m.

60 m | Car Park

100 m

16) What is the bearing of Port Q from Port P?

17) The map on the right has a scale of 1 cm = 10 miles. A ship is 15 miles from Port P on a bearing of 230° from Port P. Mark the ship on the map.

N↑

Port P

Port Q

Speed (p99) ☑

18) Write down the formula triangle linking speed, distance and time.

19) Sam swims at a speed of 1.2 m/s for 20s. How far does she swim?

Collecting Data

Data is just information. Before you collect any data, you need to think carefully about where to get it from.

Choose Your Sample Carefully

1) For any statistical project, you need to find out about a group of people or things. E.g. all the pupils in a school, or all the trees in a forest. This whole group is called the **POPULATION**.

2) Information can be collected by doing a survey — you can record observations yourself, or ask people to fill in a questionnaire.

3) Often you can't survey the whole population, e.g. because it's too big. So you select a smaller group from the population, called a **SAMPLE**, instead.

4) It's really important that your sample fairly represents the WHOLE population. This allows you to apply any conclusions from your survey to the whole population.

You Need to Spot Problems with Sampling Methods

A BIASED sample (or survey) is one that doesn't properly represent the whole population.

To SPOT BIAS, you need to think about:

> 1) WHEN, WHERE and HOW the sample is taken.
> 2) HOW MANY members are in it.

• If any groups are left out of the sample, there can be BIAS in things like age, gender, or different interests.

• If the sample is too small, it's also likely to be biased.   *Bigger populations need bigger samples to represent them.*

If possible, the best way to AVOID BIAS is to select a large sample at random from the whole population.

EXAMPLE:
Tina wants to find out how often people travel by train.
She decides to ask the people waiting for trains at her local train station one morning.
Give one reason why this might not be a suitable sample to choose.

The sample is biased because there won't be anyone who never uses the train and there will probably be a lot of people who use the train regularly.

> Think about when, where and how Tina selects her sample.

Or you could say that the sample is only taken at one particular place and time, so won't represent everyone.

EXAMPLE:
Samir's school has 800 pupils. Samir is interested in whether these pupils would like to have more music lessons. For his sample he selects 10 members of the school orchestra.

a) Explain why Samir's sample is likely to be biased.

Only members of the orchestra are included, so it's likely to be biased in favour of more music lessons. And a sample of 10 is too small to represent the whole school.

> Think about how and how many.

b) Suggest how Samir could improve his sampling method.

Samir should choose a larger sample and randomly select pupils from the whole school.

> Think about how he could make the sample more representative of all 800 pupils.

If you ask me, I love this page — but I'm biased...

Make sure you know how to spot poor sampling methods, then take on this Exam Practice Question.

Q1   An investigation into the average number of people in households in Britain was done by surveying 100 households in one city centre.
Give two reasons why this is a poor sampling technique.   [2 marks]

Collecting Data

<u>Revision tip</u>: take regular dancing breaks — try a quick waltz spin or salsa shimmy... then back to the maths...

You Can Record Your Data in a Table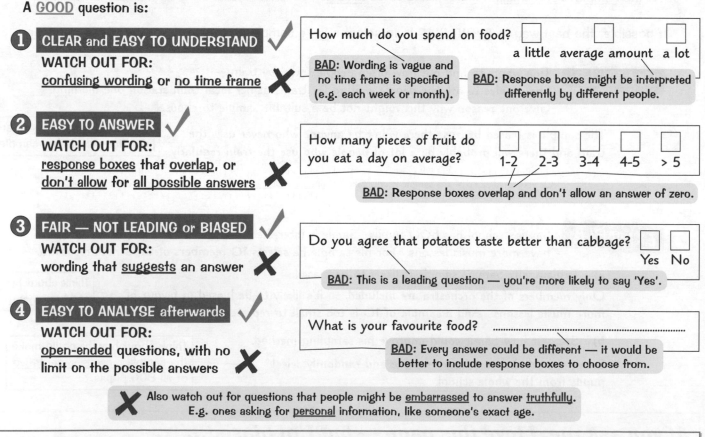

1) <u>Data-collection tables</u> (or <u>sheets</u>) should look like this table.
2) The <u>first column</u> can contain <u>words</u> (like the table opposite) or <u>numbers</u>. Make sure you include a <u>category</u> to fit <u>every possible</u> data label or value.
3) Tables for <u>grouped</u> data are covered on page 110.

Record how many

Data labels or values

Vehicle	Tally	Frequency
Car	┼┼┼ ‖	7
Bus	‖‖	4
Lorry	‖‖	3
Other	‖‖	4

Find totals by adding tally marks

EXAMPLE: Lizzie plans to ask people how many holidays they have been on this year. Design a data-collection sheet she could use to collect her data.

❶ Draw and label the <u>3 columns</u>.
The <u>data values</u> are the <u>numbers of holidays</u>.

❷ Make sure there's a place to record <u>all the answers</u> Lizzie might get. The category '<u>4 or more</u>' is a good way of doing this without having to add rows for 5, 6, 7, etc.

Number of holidays	Tally	Frequency
0		
1		
2		
3		
4 or more		

Design Your Questionnaire Carefully

You need to be able to <u>say what's wrong</u> with questionnaire <u>questions</u> and <u>write</u> your own <u>good questions</u>.

A <u>GOOD</u> question is:

❶ **CLEAR and EASY TO UNDERSTAND** ✓
WATCH OUT FOR:
<u>confusing wording</u> or <u>no time frame</u> ✗

How much do you spend on food? ☐ ☐ ☐
a little average amount a lot

BAD: Wording is vague and no time frame is specified (e.g. each week or month).

BAD: Response boxes might be interpreted differently by different people.

❷ **EASY TO ANSWER** ✓
WATCH OUT FOR:
<u>response boxes</u> that <u>overlap</u>, or <u>don't allow</u> for <u>all possible answers</u> ✗

How many pieces of fruit do you eat a day on average? ☐ ☐ ☐ ☐ ☐
1-2 2-3 3-4 4-5 > 5

BAD: Response boxes overlap and don't allow an answer of zero.

❸ **FAIR — NOT LEADING or BIASED** ✓
WATCH OUT FOR:
wording that <u>suggests</u> an answer ✗

Do you agree that potatoes taste better than cabbage? ☐ ☐
Yes No

BAD: This is a leading question — you're more likely to say 'Yes'.

❹ **EASY TO ANALYSE afterwards** ✓
WATCH OUT FOR:
<u>open-ended</u> questions, with no limit on the possible answers ✗

What is your favourite food? ....

BAD: Every answer could be different — it would be better to include response boxes to choose from.

✗ Also watch out for questions that people might be <u>embarrassed</u> to answer <u>truthfully</u>. E.g. ones asking for <u>personal</u> information, like someone's exact age.

Who wants to collect a questionnaire — the (not so exciting) quiz spin-off...

Make sure you learn the 4 key points for writing good questions, then try this Exam Practice Question.

Q1 The four questions on the page above are to be included on a questionnaire about food. Design a better version of each question to go on the questionnaire. **[4 marks]** ©

Mean, Median, Mode and Range

Mean, median, mode and range pop up all the time in statistics questions — make sure you know what they are.

MODE = MOST common (F)

MEDIAN = MIDDLE value (when values are in order of size)

MEAN = TOTAL of items ÷ NUMBER of items

RANGE = Difference between highest and lowest

REMEMBER:
Mode = most (emphasise the 'mo' in each when you say them)
Median = mid (emphasise the m*d in each when you say them)
Mean is just the average, but it's mean 'cos you have to work it out.

The Golden Rule

There's one vital step for finding the median that lots of people forget:

Always REARRANGE the data in ASCENDING ORDER
(and check you have the same number of entries!)

You must do this when finding the median, but it's also really useful for working out the mode too.

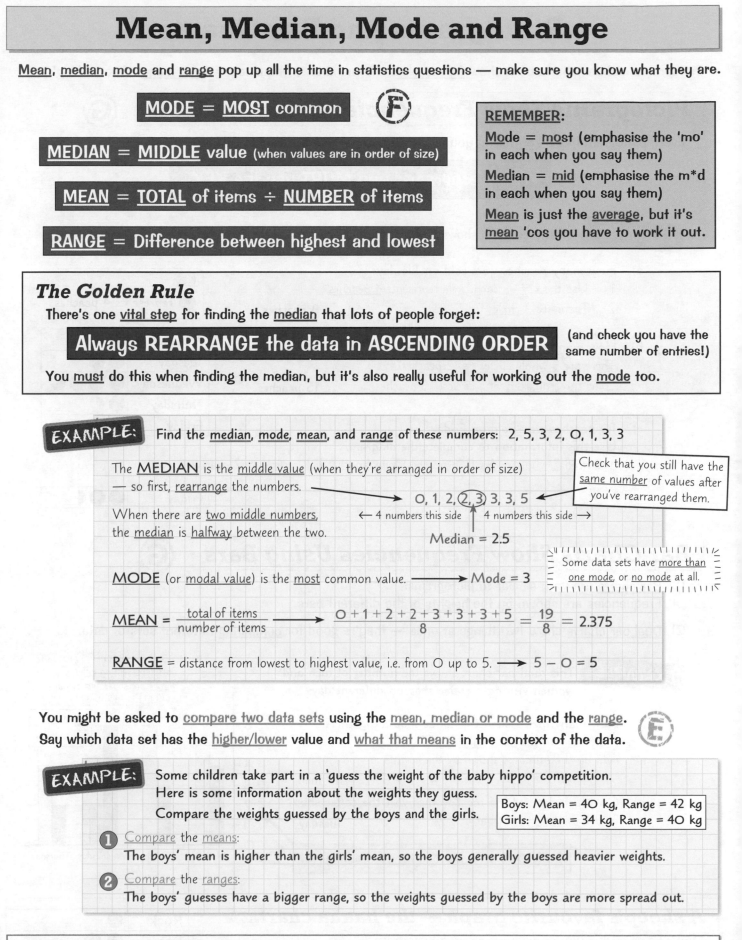

EXAMPLE: Find the median, mode, mean, and range of these numbers: 2, 5, 3, 2, 0, 1, 3, 3

The **MEDIAN** is the middle value (when they're arranged in order of size) — so first, rearrange the numbers.

Check that you still have the same number of values after you've rearranged them.

$$0, 1, 2, (2, 3) \ 3, 3, 5$$

← 4 numbers this side   4 numbers this side →

When there are two middle numbers, the median is halfway between the two.

Median = 2.5

MODE (or modal value) is the most common value. ⟶ Mode = 3

Some data sets have more than one mode, or no mode at all.

$$\text{MEAN} = \frac{\text{total of items}}{\text{number of items}} \longrightarrow \frac{0 + 1 + 2 + 2 + 3 + 3 + 3 + 5}{8} = \frac{19}{8} = 2.375$$

RANGE = distance from lowest to highest value, i.e. from 0 up to 5. ⟶ 5 − 0 = 5

You might be asked to compare two data sets using the mean, median or mode and the range. (E)
Say which data set has the higher/lower value and what that means in the context of the data.

EXAMPLE: Some children take part in a 'guess the weight of the baby hippo' competition.
Here is some information about the weights they guess.
Compare the weights guessed by the boys and the girls.

Boys: Mean = 40 kg, Range = 42 kg
Girls: Mean = 34 kg, Range = 40 kg

1 Compare the means:
The boys' mean is higher than the girls' mean, so the boys generally guessed heavier weights.

2 Compare the ranges:
The boys' guesses have a bigger range, so the weights guessed by the boys are more spread out.

My favourite range is the Alps...

Learn the four definitions, then cover the page and write them down. Then do this practice question.

Q1 a) Find the mean, median, mode and range for these test scores: 6, 15, 12, 12, 11.   [4 marks] (F)
   b) Another set of scores has a mean of 9 and a range of 12. Compare the two sets.   [2 marks] (E)

Pictograms and Bar Charts

Pictograms and bar charts both show <u>frequencies</u>. (Remember... frequency = '<u>how many</u> of something'.)

Pictograms Show Frequencies Using Symbols (G)

Every pictogram has a <u>key</u> telling you what one symbol represents.

> With pictograms, you **<u>MUST</u>** use the **<u>KEY</u>**.

EXAMPLE: This pictogram shows how many peaches were sold by a greengrocer on different days.

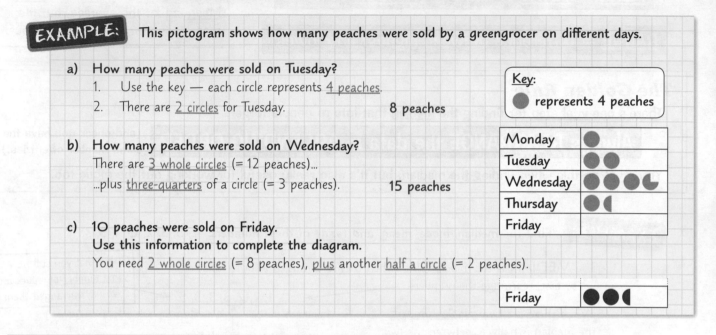

a) How many peaches were sold on Tuesday?
   1. Use the key — each circle represents <u>4 peaches</u>.
   2. There are <u>2 circles</u> for Tuesday.   **8 peaches**

b) How many peaches were sold on Wednesday?
   There are <u>3 whole circles</u> (= 12 peaches)...
   ...plus <u>three-quarters</u> of a circle (= 3 peaches).   **15 peaches**

c) 10 peaches were sold on Friday.
   Use this information to complete the diagram.
   You need <u>2 whole circles</u> (= 8 peaches), <u>plus</u> another <u>half a circle</u> (= 2 peaches).

Bar Charts Show Frequencies Using Bars (G)

1) <u>Bar charts</u> are very similar to pictograms.
   Frequencies are shown by the <u>heights</u> of the different bars.

2) <u>Dual bar charts</u> show two things at once — they're good for <u>comparing</u> different sets of data.

EXAMPLE: This dual bar chart shows the number of men and women visiting a coffee shop on different days.

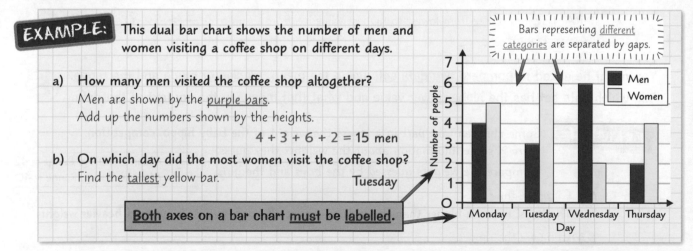

a) How many men visited the coffee shop altogether?
   Men are shown by the <u>purple bars</u>.
   Add up the numbers shown by the heights.
   $4 + 3 + 6 + 2 = 15$ men

b) On which day did the most women visit the coffee shop?
   Find the <u>tallest</u> yellow bar.   **Tuesday**

> <u>Both</u> axes on a bar chart <u>must</u> be <u>labelled</u>.

A sheep's favourite graph — the baaaa chart...

Don't be scared if you're asked to draw a histogram —
it's just a bar chart that you draw using a frequency table.

Q1   This pictogram shows the different types of CDs Javier owns.
   a) How many jazz CDs does Javier own?   [1 mark]
   b) He owns 5 opera CDs. Complete the pictogram.   [1 mark]

Rock	●
Blues	●●
Opera	
Jazz	●◖

<u>Key:</u> ● represents 2 CDs

Two-Way Tables

Two-way tables are another way to represent data.
They show how many things or people fall into different categories.

Two-Way Tables Show Frequencies (E)

Fill in any information you're <u>not</u> given in the question by <u>adding</u> or <u>subtracting</u>.

EXAMPLE: This two-way table shows the number of cakes and loaves of bread a bakery sells on Friday and Saturday one week.

	Cakes	Loaves of bread	Total
Friday		10	22
Saturday	4	14	
Total	16	24	40

a) Work out how many items in total were sold on Saturday.

<u>Either</u>: (i) <u>add</u> the number of <u>cakes</u> for Saturday to the number of <u>loaves of bread</u>.

<u>Or</u>: (ii) <u>take away</u> the total items sold on <u>Friday</u> from the total sold over <u>both days</u>.

	Cakes	Loaves of bread	Total
Friday		10	22
Saturday	4	14	18
Total	16	24	40

b) Work out how many cakes were sold on Friday.

<u>Either</u>: (i) <u>take away</u> the <u>loaves of bread</u> for Friday from the <u>total</u> number of items for Friday.

<u>Or</u>: (ii) <u>take away</u> the number of cakes for <u>Saturday</u> from the total number of cakes over <u>both days</u>.

	Cakes	Loaves of bread	Total
Friday	12	10	22
Saturday	4	14	18
Total	16	24	40

You Might Have to Draw Your Own Table (E)

Sometimes they don't even give you a table — just <u>a few bits</u> of information.

EXAMPLE: 200 men and 200 women are asked whether they are left-handed or right-handed.
- 63 people altogether are left-handed.
- 164 of the women are right-handed.

When there's only <u>one</u> thing in a row or column that you don't know, you can <u>always</u> work it out.

How many of the men are right-handed?

1. Draw a table to show the information from the <u>question</u> — this is in the <u>yellow cells</u>.
2. Then fill in the gaps by <u>adding</u> and <u>subtracting</u>.

	Women	Men	Total
Left-handed	200 − 164 = 36	63 − 36 = 27	63
Right-handed	164	200 − 27 = 173	400 − 63 = 337
Total	200	200	200 + 200 = 400

So there are 173 right-handed men.

If you like sudoku, you'll love two-way tables...

This exam question has a bigger two-way table — with 5 columns instead of 4. Complete it by adding and subtracting — just like always.

Q1 Complete the two-way table on the right showing how a group of students get to school. [3 marks] (E)

	Walk	Car	Bus	Total
Male	15	21		
Female			22	51
Total	33	32		100

Pie Charts

Just like you can make the correct combination of meat and pastry into a delicious pie, examiners can make <u>pie charts</u> into tricky exam questions. Just remember the <u>Golden Pie Chart Rule</u>...

The TOTAL of Everything = 360°

1) Fraction of the Total = Angle ÷ 360°

EXAMPLE: This pie chart shows the colour of all the cars sold by a dealer. What fraction of the cars were red?

Just remember that 'everything = 360°'.

$$\text{Fraction of red cars} = \frac{\text{angle of red cars}}{\text{angle of everything}} = \frac{72°}{360°} = \frac{1}{5}$$

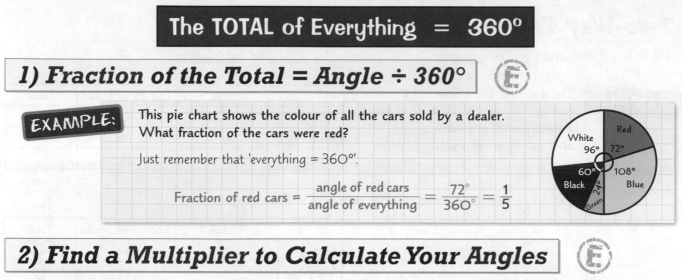

2) Find a Multiplier to Calculate Your Angles

EXAMPLE: Draw a pie chart to show this information about the types of animal in a petting zoo.

Animal	Geese	Hamsters	Guinea pigs	Rabbits	Ducks
Number	12	20	17	15	26

1. Find the <u>total</u> by <u>adding</u>.   12 + 20 + 17 + 15 + 26 = 90

2. 'Everything = 360°' — so find the <u>multiplier</u> (or <u>divider</u>) that turns your total into 360°.   Multiplier = 360 ÷ 90 = 4

3. Now <u>multiply every number</u> by 4 to get the <u>angle</u> for each sector.

Angle	12 × 4 = 48°	20 × 4 = 80°	17 × 4 = 68°	15 × 4 = 60°	26 × 4 = 104°	Total = 360°

4. Draw your pie chart accurately using a <u>protractor</u>.

3) Find How Many by Using the Angle for 1 Thing

EXAMPLE: The pie chart on the right shows information about the types of animals liked most by different students. There were 9 students altogether.

a) Work out the number of students who liked dogs most.

1. 'Everything = 360°', so...   ⟹ 9 students = 360°

2. <u>Divide by 9</u> to find...   ⟹ 1 student = 40°

3. The <u>angle</u> for dogs is 160°, so you need to <u>multiply both sides by 4</u>: ⟹ 4 students = 160° — 4 students liked dogs most

b) The pie chart on the left shows information about the types of animals liked most by a different group of students. Dave says, "This means that 4 students in this group like dogs most." Explain why Dave is not correct.

We don't know how many students in total the pie chart represents, so we can't work out how many students liked dogs most.

I like my pie charts with gravy and mushy peas...

Pie chart questions need a lot of practice. Make a start with this...

Q1   Draw an accurate pie chart to show the information about Rahul's DVD collection in this table.   [3 marks]

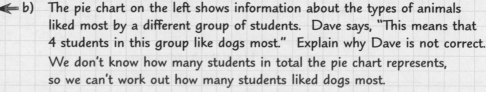

Type of DVD	Number
Rom Com	23
Western	25
Action	12

Section Seven — Statistics and Probability

Scatter Graphs

Scatter graphs are really useful — they show you if there's a <u>link</u> between two things.

Scatter Graphs Show Correlations Ⓓ

1) A <u>scatter graph</u> shows how closely two things are <u>related</u>. The fancy word for this is <u>CORRELATION</u>.

2) If the two things <u>are related</u>, then you'd be able to draw a <u>straight line</u> ← Called a <u>line of best fit</u>.
    passing <u>pretty close</u> to <u>most</u> of the points on the scatter diagram.

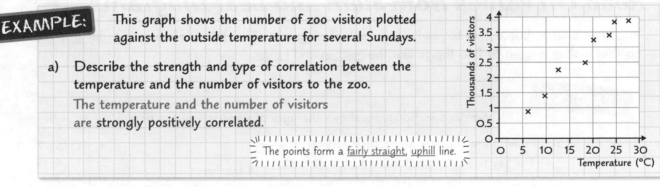

<u>STRONG correlation</u> is when your points make a <u>fairly straight line</u>.

STRONG POSITIVE CORRELATION

If the points form a line sloping <u>uphill</u> from left to right, then there is <u>POSITIVE</u> correlation — both things increase or decrease <u>together</u>.

<u>WEAK correlation</u> means your points <u>don't line up</u> quite so nicely (but you can still draw a line of best fit through them).

WEAK NEGATIVE CORRELATION

If the points form a line sloping <u>downhill</u> from left to right, then there is <u>NEGATIVE</u> correlation — as one quantity <u>increases</u>, the other <u>decreases</u>.

3) If the two things are <u>not related</u>, you get a load of <u>messy points</u>. This scatter graph is a messy scatter — so there's <u>no correlation</u> between the two things.

NO CORRELATION

EXAMPLE: This graph shows the number of zoo visitors plotted against the outside temperature for several Sundays.

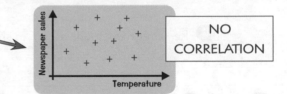

a) Describe the strength and type of correlation between the temperature and the number of visitors to the zoo.

The temperature and the number of visitors are strongly positively correlated.

The points form a <u>fairly straight</u>, <u>uphill</u> line.

4) You can use a line of best fit to <u>predict</u> other values.

b) Estimate how many visitors the above zoo would get on a Sunday when the outside temperature is 15 °C.

1. Draw a <u>line of best fit</u> (shown in <u>blue</u>).

2. Draw a line <u>up from 15 °C</u> to your line, and then <u>across to the other axis</u>.

15 °C corresponds to roughly 2250 visitors.

Relax and take a trip down Correlation Street...

You need to feel at home with scatter graphs. See how you feel with this question. Ⓓ

Q1 This graph shows Sam's average speed on runs of different lengths.

a) Describe the relationship between 'length of run' and 'average speed'. [1 mark]

b) Estimate Sam's average speed for an 8-mile run. [1 mark]

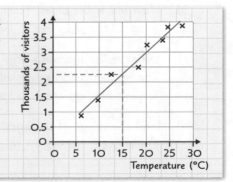

Section Seven — Statistics and Probability

Stem and Leaf Diagrams

Stem and leaf diagrams don't look as pretty as you might expect, but they're a useful way to show data.

Stem and Leaf Diagrams Put the Data in Order Ⓓ

An ordered stem and leaf diagram shows a set of data in order of size.

To draw one, you need to decide how to make the stem, then the rest isn't too bad.

EXAMPLE:

Here are the scores for 15 dogs in an agility test:

26, 16, 29, 7, 12, 32, 29, 24, 13, 17, 20, 23, 24, 31, 34

Draw an ordered stem and leaf diagram to show the data.

Check you've got the right number of values.

① First, list the data IN ORDER of size. ⟶ 7, 12, 13, 16, 17, 20, 23, 24, 24, 26, 29, 29, 31, 32, 34

② Choose numbers to put in the STEM. Here you can split the values into numbers of tens and units.
So the stem is the 'number of tens':

Stem	Leaves
no tens (single digits) ⟶ 0	7
1 ten (between 10 and 20) ⟶ 1	2 3 6 7
2 tens (20 something) ⟶ 2	0 3 4 4 6 9 9
3 tens (30 something) ⟶ 3	1 2 4

③ The LEAVES are the numbers of units. Go through the values in order and write the units in the correct row.

Key: 2|3 = 23

④ Remember to include a KEY to show how to read the diagram. E.g. 2 tens and 3 units means 23.

Read Off Values from Stem and Leaf Diagrams Ⓔ

You also need to know how to read stem and leaf diagrams. The data is already in order of size, so that makes it easy to find things like the median and range.

See p103 for a reminder about median and range.

EXAMPLE:

This stem and leaf diagram shows the ages of some school teachers.

3	3 5
4	0 5 7 8
5	1 4 9
6	1 3

Key: 5|4 = 54 years

a) **How many teachers are in their forties?**
The key tells you that 4 in the stem means 40, so count the leaves in the second row.

4 teachers

b) **How old is the oldest teacher?**
Read off the last value.

6 | 3 = 63 years old

c) **What is the median age?**
The median is the middle value.
Find its position, then read off the value.

There are 11 values, so the median is the 6th value.
So median age = 48 years

Look at the stem and the leaf — it's 48, not 8.

d) **Find the range of the ages.**
The range is the highest minus the lowest.

Range = 63 − 33 = 30 years

From a tiny seed of maths grows a beautiful stem and leaf diagram...

Learn how to draw and read stem and leaf diagrams. Then do this Exam Practice Question.

Q1 a) Draw an ordered stem and leaf diagram to show: 3, 16, 14, 22, 7, 11, 26, 17, 12   [3 marks] Ⓓ
    b) Use your diagram to find the median and range of the data.   [3 marks] Ⓔ

Frequency Tables and Averages

The word <u>FREQUENCY</u> means <u>HOW MANY</u>, so a frequency table is just a '<u>How many in each category' table</u>.

E.g. in the table opposite, 17 people don't have a cat, **22** have one cat, etc.

You found <u>averages</u> on page 103, and you can do the same sort of thing from tables.

How many

Categories

Number of cats	Frequency	
0	17	
1	22	
2	15	
3	7	

Mysterious 3rd column...

> 1) The <u>MODE</u> is just the <u>CATEGORY</u> with the <u>MOST ENTRIES</u>.
>
> 2) The <u>MEDIAN</u> is the <u>CATEGORY</u> of the <u>middle value</u>.
>
> 3) To find the <u>MEAN</u>, you have to <u>WORK OUT A THIRD COLUMN</u> yourself.
>
> The <u>MEAN</u> is then: <u>3rd Column Total ÷ 2nd Column Total</u>

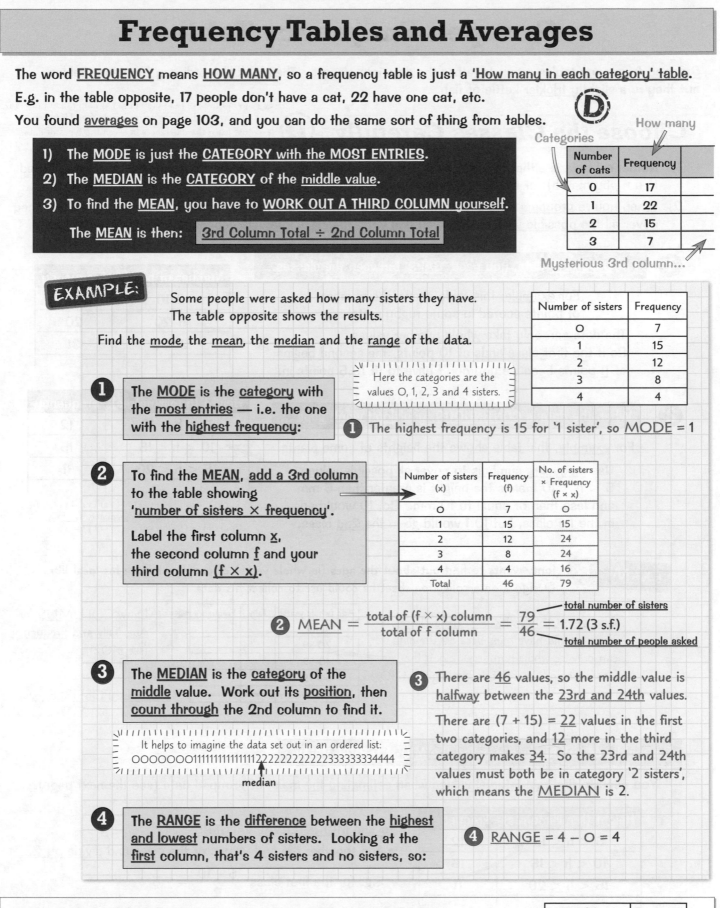

EXAMPLE:

Some people were asked how many sisters they have. The table opposite shows the results.

Find the <u>mode</u>, the <u>mean</u>, the <u>median</u> and the <u>range</u> of the data.

Number of sisters	Frequency
0	7
1	15
2	12
3	8
4	4

1 The <u>MODE</u> is the <u>category</u> with the <u>most entries</u> — i.e. the one with the <u>highest frequency</u>:

Here the categories are the values 0, 1, 2, 3 and 4 sisters.

1 The highest frequency is 15 for '1 sister', so <u>MODE</u> = 1

2 To find the <u>MEAN</u>, <u>add a 3rd column</u> to the table showing 'number of sisters × frequency'.

Label the first column <u>x</u>, the second column <u>f</u> and your third column <u>(f × x)</u>.

Number of sisters (x)	Frequency (f)	No. of sisters × Frequency (f × x)
0	7	0
1	15	15
2	12	24
3	8	24
4	4	16
Total	46	79

total number of sisters

2 $\underline{\text{MEAN}} = \dfrac{\text{total of (f × x) column}}{\text{total of f column}} = \dfrac{79}{46} = 1.72 \text{ (3 s.f.)}$

total number of people asked

3 The <u>MEDIAN</u> is the <u>category</u> of the <u>middle</u> value. Work out its <u>position</u>, then <u>count through</u> the 2nd column to find it.

It helps to imagine the data set out in an ordered list:
00000001111111111111112222222222222333333334444
↑
median

3 There are <u>46</u> values, so the middle value is <u>halfway</u> between the <u>23rd and 24th</u> values.

There are (7 + 15) = <u>22</u> values in the first two categories, and <u>12</u> more in the third category makes <u>34</u>. So the 23rd and 24th values must both be in category '2 sisters', which means the <u>MEDIAN</u> is 2.

4 The <u>RANGE</u> is the <u>difference</u> between the <u>highest and lowest</u> numbers of sisters. Looking at the <u>first</u> column, that's 4 sisters and no sisters, so:

4 <u>RANGE</u> = 4 − 0 = 4

My table has 5 columns, 6 rows and 4 legs...

Learn the three key points about averages, then try this Exam Practice Question.

Q1   50 people were asked how many times a week they play sport. The table opposite shows the results.
  a) Find the median.   [2 marks]
  b) Calculate the mean.   [3 marks]

No. of times sport played	Frequency
0	8
1	15
2	17
3	6
4	4
5 or more	0

Grouped Frequency Tables

Grouped frequency tables group together the data into classes. They look like ordinary frequency tables, but they're a slightly trickier kettle of fish...

Choose the Classes Carefully Ⓓ

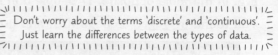
Don't worry about the terms 'discrete' and 'continuous'. Just learn the differences between the types of data.

1) Numerical data is either <u>discrete</u> — it can only take <u>certain exact values</u> (e.g. number of points scored in a rugby match), or it's <u>continuous</u> — it can take <u>any value</u> in a range (e.g. height of a plant).

2) When you're <u>grouping</u> data in a table, you need to make sure that <u>none of the classes overlap</u> and they <u>cover all the possible values</u>. The way you <u>write the class intervals</u> depends on the <u>type of data</u>...

1 For <u>DISCRETE</u> data, you need 'GAPS' between the classes...

For <u>example</u>, this table shows the number of points scored in some rugby matches.

The data can only take <u>whole number</u> values. So if the <u>first class ends</u> at <u>10</u> points, the <u>second begins</u> at <u>11</u> points because you don't have to fit 10.5 points in.

Number of points	Frequency
0-10	5
11-20	20
21-30	31

2 When the data is <u>CONTINUOUS</u>, you <u>don't</u> want 'gaps'...

For <u>example</u>, this table shows the heights of some plants.

Use <u>inequality symbols</u> to <u>cover all possible values</u>. $5 < h \leq 10$ means the height is greater than 5 mm and less than or equal to 10 mm. So, <u>10</u> would go in the <u>1st</u> class, but <u>10.1</u> would go in the <u>2nd</u> class.

Height (h millimetres)	Frequency
$5 < h \leq 10$	12
$10 < h \leq 15$	15
$15 < h \leq 20$	11

See p44 for more on <u>inequality symbols</u>.

EXAMPLE: Jonty wants to find out about the ages (in whole years) of people who use his local library. Design a data-collection sheet he could use to collect his data.

<u>Choose NON-OVERLAPPING</u> classes. These should have <u>gaps</u> because the data can only be <u>whole numbers</u>. E.g. if the 2nd class ends at 39, the 3rd begins at 40.

Age (whole years)	Tally	Frequency
0-19		
20-39		
40-59		
60-79		
80 or over		

Draw <u>COLUMNS</u> for <u>Age</u>, <u>Tally</u> and <u>Frequency</u> (see p102).

<u>Cover ALL POSSIBLE</u> values.

The <u>middle</u> of a data <u>class</u> is called the <u>MID-INTERVAL VALUE</u>.

You need to <u>find mid-interval values</u> when <u>estimating the mean</u> of grouped data (see the next page).

Height (h millimetres)	Frequency
$5 < h \leq 10$	12
$10 < h \leq 15$	15
$15 < h \leq 20$	11

<u>To find MID-INTERVAL VALUES:</u>
• Add together the <u>end values</u> of the <u>class</u> and <u>divide by 2</u>.
• E.g. for the first class: $\dfrac{5 + 10}{2} = \underline{7.5}$

Mid-interval value — cheap ice creams...

Learn all the details on the page, then try this Exam Practice Question.

Q1 Here are the heights of some adults to the nearest 0.1 cm. Design and fill in a frequency table to record the data. [3 marks]

150.4	163.5	156.7	164.1
182.8	175.4	171.2	169.0
173.3	185.6	167.0	162.6

Ⓓ

Grouped Frequency Tables — Averages

Another page on grouped frequency tables... lucky you. The methods for finding averages are similar to the ones used for normal frequency tables (see p109), except now you need to add two extra columns.

Find Averages from Grouped Frequency Tables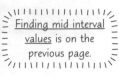

Unlike with ordinary frequency tables, you don't know the actual data values, only the classes they're in. So you have to ESTIMATE THE MEAN, rather than calculate it exactly. Again, you do this by adding columns:

Add a 3RD COLUMN and enter the MID-INTERVAL VALUES for each class.

Add a 4TH COLUMN to show 'FREQUENCY × MID-INTERVAL VALUE' for each class.

The ESTIMATED MEAN is then: | 4th Column Total ÷ 2nd Column Total |

Finding mid interval values is on the previous page.

And you'll be asked to find the MODAL CLASS and the CLASS CONTAINING THE MEDIAN, not exact values.

EXAMPLE:

This table shows information about the weights, in kilograms, of 60 schoolchildren.

a) Write down the modal class.
b) Write down the class containing the median.
c) Calculate an estimate for the mean weight.

Weight (w kg)	Frequency
30 < w ≤ 40	8
40 < w ≤ 50	16
50 < w ≤ 60	18
60 < w ≤ 70	12
70 < w ≤ 80	6

a) The modal class is the one with the highest frequency. → Modal class is 50 < w ≤ 60

b) Work out the position of the median, then count through the 2nd column.

There are 60 values, so the median is halfway between the 30th and 31st values.

Both these values are in the third class, so the class containing the median is 50 < w ≤ 60.

c) Add extra columns for 'mid-interval value' and 'frequency × mid-interval value'.

Label the frequency column f, your third column x and your fourth column (f × x).

Weight (w kg)	Frequency (f)	Mid-interval value (x)	(f × x)
30 < w ≤ 40	8	35	280
40 < w ≤ 50	16	45	720
50 < w ≤ 60	18	55	990
60 < w ≤ 70	12	65	780
70 < w ≤ 80	6	75	450
Total	60	—	3220

The mid-interval values are used to work out the 4th column. You don't need to add them up.

Mean = $\dfrac{\text{total of (f × x) column}}{\text{total of f column}}$ ← 4th column total
← 2nd column total

$= \dfrac{3220}{60}$

= 53.7 kg (3 s.f.)

This page auditioned for Britain's next top modal...

Learn all the details about the mean, median and modal class, then have a go at this practice question.

Q1 Estimate the mean of this data. Give your answer to 3 significant figures. [4 marks]

Length (l cm)	15.5 ≤ l < 16.5	16.5 ≤ l < 17.5	17.5 ≤ l < 18.5	18.5 ≤ l < 19.5
Frequency	12	18	23	8

Frequency Polygons

Frequency polygons can be used to show data from grouped frequency tables.

Frequency Polygons Show Frequencies (D)

Frequency polygons can be <u>fiddly</u> to draw. But remember this <u>rule</u> and you'll be okay.

Always plot your point at the <u>mid-interval value</u> of a class.

EXAMPLE: Draw a frequency polygon to show the information in this table.

Age (a) of people at concert	Frequency	mid-interval value
$10 < a \leq 20$	20	$(10 + 20) \div 2 = \underline{15}$
$20 < a \leq 30$	36	$(20 + 30) \div 2 = \underline{25}$
$30 < a \leq 40$	32	$(30 + 40) \div 2 = \underline{35}$
$40 < a \leq 50$	26	$(40 + 50) \div 2 = \underline{45}$
$50 < a \leq 60$	24	$(50 + 60) \div 2 = \underline{55}$

Add the endpoints of the class, then <u>divide by 2</u>.

1. Add a column to show the <u>mid-interval values</u>.

2. Plot the <u>mid-interval values</u> on the <u>horizontal axis</u> and the <u>frequencies</u> on the <u>vertical axis</u>.

   So plot the points (<u>15, 20</u>), (<u>25, 36</u>), (<u>35, 32</u>), (<u>45, 26</u>) and (<u>55, 24</u>).

3. <u>Join</u> the points of your frequency polygon using <u>straight lines</u> (not a curve).

Use Frequency Polygons to Interpret Sets of Data (D)

EXAMPLE: The frequency polygon <u>in blue</u> represents the ages of people at a football game (while the <u>red</u> frequency polygon shows the ages of people at the above concert).

Make one comment to compare the two distributions.

Whatever you say must involve <u>both</u> distributions. There were more people over 40 years old at the football game than at the concert.

There are loads of things you could say. For example, you could say there were far more people aged 20-30 at the concert than at the football game.

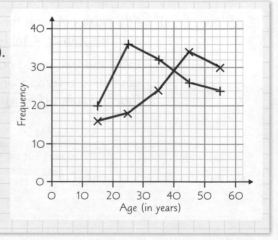

Practise these polygons frequently...

If you learn the stuff above, then you should be able to cope with any question you get in the exam. And with this question too.

Q1 Draw a frequency polygon to represent the information shown in this frequency table. **[2 marks]** (D)

Height (h, in m) of student	Frequency
$1.0 < h \leq 1.2$	5
$1.2 < h \leq 1.4$	8
$1.4 < h \leq 1.6$	6
$1.6 < h \leq 1.8$	2

Probability Basics

A lot of people think <u>probability</u> is tough. But learn the <u>basics</u> well, and it'll all make sense.

All Probabilities are Between 0 and 1 (F)

- Probabilities are <u>always</u> between 0 and 1.
- The <u>higher</u> the probability of something, the <u>more likely</u> it is.
- A probability of <u>ZERO</u> means it will <u>NEVER HAPPEN</u>.
- A probability of <u>ONE</u> means it <u>DEFINITELY WILL HAPPEN</u>.

You can show the probability of something happening on a <u>scale</u> from 0 to 1.
Probabilities can be given as either <u>fractions</u> or <u>decimals</u>.

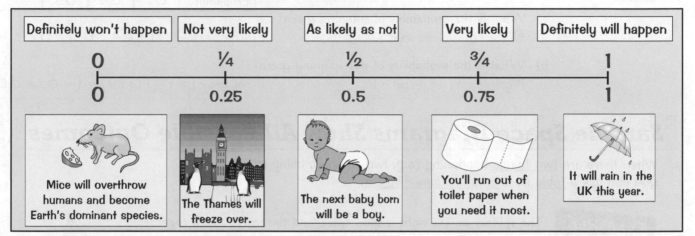

Definitely won't happen	Not very likely	As likely as not	Very likely	Definitely will happen
0	¼	½	¾	1
0	0.25	0.5	0.75	1

Mice will overthrow humans and become Earth's dominant species.
The Thames will freeze over.
The next baby born will be a boy.
You'll run out of toilet paper when you need it most.
It will rain in the UK this year.

Use This Formula When All Outcomes are Equally Likely (E)

Use this formula to find probabilities for a <u>fair</u> spinner, coin or dice.
A spinner/coin/dice is 'fair' when it's <u>equally likely</u> to land on <u>any</u> of its sides.

$$\text{Probability} = \frac{\text{Number of ways for something to happen}}{\text{Total number of possible outcomes}}$$

<u>Outcomes</u> are just 'things that could happen'.

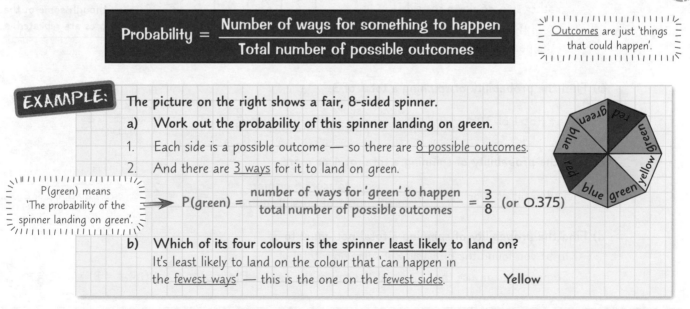

EXAMPLE: The picture on the right shows a fair, 8-sided spinner.

a) Work out the probability of this spinner landing on green.

1. Each side is a possible outcome — so there are <u>8 possible outcomes</u>.
2. And there are <u>3 ways</u> for it to land on green.

P(green) means 'The probability of the spinner landing on green'.

$$\text{P(green)} = \frac{\text{number of ways for 'green' to happen}}{\text{total number of possible outcomes}} = \frac{3}{8} \text{ (or 0.375)}$$

b) Which of its four colours is the spinner <u>least likely</u> to land on?

It's least likely to land on the colour that 'can happen in the <u>fewest ways</u>' — this is the one on the <u>fewest sides</u>. **Yellow**

The probability of this getting you marks in the exam = 1...

You need to know the facts in the boxes above. You also need to know how to <u>use</u> them. (E)

Q1   a) Calculate the probability of the spinner on the right landing on 4.   [2 marks]

       b) Show this probability on a scale from 0 to 1.   [1 mark]

More Probability

The formulas in the boxes below are probably the <u>most important</u> probability formulas ever. Use them wisely.

Probabilities Add Up To 1 (D)

1) If <u>only one</u> possible result can happen at a time, then the probabilities of <u>all</u> the results <u>add up to 1</u>.

Probabilities always ADD UP to 1.

2) So since something must either <u>happen</u> or <u>not happen</u> (i.e. <u>only one</u> of these can happen at a time):

P(event happens) + P(event doesn't happen) = 1

EXAMPLE: A spinner has different numbers of red, blue, yellow and green sections.

Colour	red	blue	yellow	green
Probability	0.1	0.4	0.3	

a) What is the probability of spinning green?
All the probabilities must <u>add up to 1</u>.   P(green) = 1 − (0.1 + 0.4 + 0.3) = 0.2

b) What is the probability of <u>not</u> spinning green?
P(green) + P(not green) = 1   P(not green) = 1 − P(green) = 1 − 0.2 = 0.8

Sample Space Diagrams Show All Possible Outcomes (D)

When there are <u>two things</u> happening (e.g. two spinners being spun),
you can use a <u>table</u> as a <u>sample space diagram</u>.

EXAMPLE: The spinner on the right is spun twice, and the scores added together.

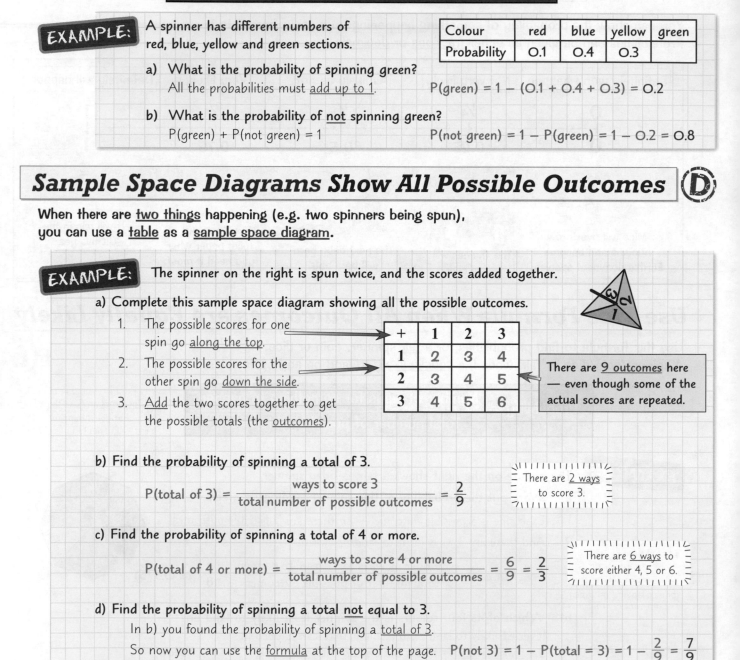

a) Complete this sample space diagram showing all the possible outcomes.

1. The possible scores for one
    spin go <u>along the top</u>.

2. The possible scores for the
    other spin go <u>down the side</u>.

3. <u>Add</u> the two scores together to get
    the possible totals (the <u>outcomes</u>).

+	1	2	3
1	2	3	4
2	3	4	5
3	4	5	6

There are <u>9 outcomes</u> here — even though some of the actual scores are repeated.

b) Find the probability of spinning a total of 3.

$$P(\text{total of 3}) = \frac{\text{ways to score 3}}{\text{total number of possible outcomes}} = \frac{2}{9}$$

There are <u>2 ways</u> to score 3.

c) Find the probability of spinning a total of 4 or more.

$$P(\text{total of 4 or more}) = \frac{\text{ways to score 4 or more}}{\text{total number of possible outcomes}} = \frac{6}{9} = \frac{2}{3}$$

There are <u>6 ways</u> to score either 4, 5 or 6.

d) Find the probability of spinning a total <u>not</u> equal to 3.

In b) you found the probability of spinning a <u>total of 3</u>.
So now you can use the <u>formula</u> at the top of the page. $P(\text{not 3}) = 1 - P(\text{total = 3}) = 1 - \frac{2}{9} = \frac{7}{9}$

When in doubt, make a list...

If you're ready to put your statistics skills to the test, try this...

Q1   Two fair dice are thrown, and their scores added together. (D)
  a) Find the probability of throwing a total of 7.   [2 marks]
  b) Find the probability of throwing any total <u>except 7</u>.   [1 mark]

Section Seven — Statistics and Probability

Expected Frequency

You can use probabilities to work out how often you'd expect something to happen.

Use Probability to Find an "Expected Frequency" Ⓓ

1) Once you know the probability of something, you can predict how many times it will happen in a certain number of trials. *A 'trial' could be any activity — e.g. rolling a dice.*

2) For example, you can predict the number of sixes you could expect if you rolled a fair dice 20 times. This prediction is called the expected frequency.

Expected frequency = probability × number of trials

EXAMPLE:
The probability of someone catching a frisbee thrown to them is 0.92.
Estimate the number of times you would expect them to catch a frisbee in 150 attempts.

Expected number of catches = probability of a catch × number of trials

= 0.92 × 150

= 138

This is an estimate. They might not catch the frisbee exactly 138 times, but the number of catches shouldn't be too different from this.

You Might Have to Find a Probability First Ⓓ

EXAMPLES:

1. A person spins the fair spinner on the right 200 times.
How many times would you expect it to land on 5?

1. First calculate the probability of the spinner landing on 5.

$$P(\text{lands on 5}) = \frac{\text{ways to land on 5}}{\text{number of possible outcomes}} = \frac{1}{8}$$

2. Then estimate the number of 5's they'll get in 200 spins.

Expected number of 5's = P(lands on 5) × number of trials

$$= \frac{1}{8} \times 200 = 25$$

2. I buy 400 large tins of chocolates.
Each tin contains 100 chocolates altogether, and 80 of these are milk chocolate.
If I select one chocolate at random from each tin, how many milk chocolates would I expect to get?

1. First calculate the probability of picking a milk chocolate from one tin.

$$P(\text{milk chocolate from 1 tin}) = \frac{\text{number of ways to get a milk chocolate}}{\text{total number of chocolates in each tin}}$$

$$= \frac{80}{100} = \frac{4}{5}$$

2. Then estimate the number of milk chocolates if I pick one chocolate from each of the 400 tins.

Expected milk chocolates = P(milk chocolate from 1 tin) × number of tins

$$= \frac{4}{5} \times 400 = 320$$

I predict this page could earn you 3 marks in your exam...

This is why statistics is so cool — you can use it to predict the future. Ⓓ

Q1 A game involves throwing a fair dice once. The player wins if they score either a 5 or a 6.
If one person plays the game 180 times, estimate the number of times they will win. [3 marks]

Relative Frequency

This isn't the number of times your granny comes to visit. It's a way of working out probabilities.

Fair or Biased?

'Fair' just means that all the possible scores are equally likely.

1) You can use the formula on p113 to work out that the probability of rolling a 3 on a dice is $\frac{1}{6}$.
2) BUT this only works if it's a fair dice. If the dice is wonky (the technical term is 'biased') then each number won't have an equal chance of being rolled.
3) This is where relative frequency is useful. You can use it to estimate probabilities when things are wonky.

Do the Experiment Again and Again and Again...

You need to do an experiment over and over again and count how often a result happens (its frequency). Then you can find its relative frequency.

$$\text{Relative frequency} = \frac{\text{Frequency}}{\text{Number of times you tried the experiment}}$$

An experiment could just mean rolling a dice.

You can use the relative frequency of a result as an estimate of its probability.

EXAMPLE:

The spinner on the right was spun 100 times. The results are in the table below. Estimate the probability of getting each of the scores.

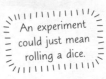

Score	1	2	3	4	5	6
Frequency	10	14	36	20	11	9

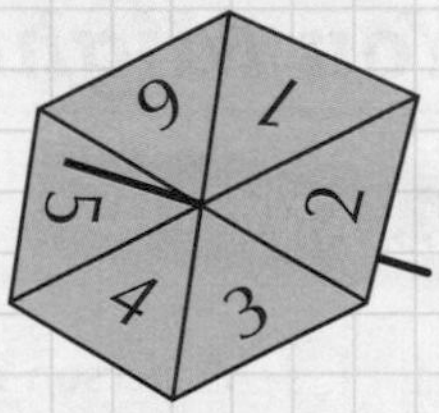

The spinner was spun 100 times.
So divide each of the frequencies by 100 to find the relative frequencies.

Score	1	2	3	4	5	6
Relative Frequency	$\frac{10}{100} = 0.1$	$\frac{14}{100} = 0.14$	$\frac{36}{100} = 0.36$	$\frac{20}{100} = 0.2$	$\frac{11}{100} = 0.11$	$\frac{9}{100} = 0.09$

The MORE TIMES you do the experiment, the MORE ACCURATE your estimate of the probability will be. If you spun the above spinner 1000 times, you'd get a better estimate of the probability of each score.

For a fair dice or spinner, the relative frequencies should all be roughly the same after a large number of trials. If some of them are very different, the dice or spinner is probably biased.

Do the above results suggest that the spinner is biased?
Yes, because the relative frequency of 3 is much higher than the relative frequencies of 1, 5 and 6.

For a fair 6-sided spinner, you'd expect all the relative frequencies to be about $1 \div 6 = 0.17$(ish).

This is a tough topic — make sure you revise it relatively frequently...

If a coin/dice/spinner is fair, then you can tell the probability of each of the results basically 'just by looking at it'. But if it's biased, then you have to use relative frequencies to estimate the probabilities.

Q1 Sandro threw a dice 1000 times and got the results shown in the table below.

Score	1	2	3	4	5	6
Frequency	140	137	138	259	161	165

a) Find the relative frequencies for each of the scores 1-6. [2 marks]
b) Do these results suggest that the dice is biased? Give a reason for your answer. [1 mark]

Revision Questions for Section Seven

Here's the inevitable list of straight-down-the-middle questions to test how much you know.
- Have a go at each question... but <u>only tick it off</u> when you can get it right <u>without</u> cheating.
- And when you think you could handle pretty much <u>any</u> statistics question, tick off the whole topic.

Collecting Data and Finding Averages (p101-103)

Pet	Tally	Frequency

1) What is a sample and why does it need to be representative?
2) Complete this frequency table for the data below. ➡
      Cat, Cat, Dog, Dog, Dog, Rabbit, Fish, Cat, Rabbit, Rabbit, Dog, Dog, Cat, Cat, Dog, Rabbit, Cat, Fish, Cat, Cat
3) List the four key things you should bear in mind when writing questionnaire questions.
4) Find the mode, median, mean and range of this data: 2, 8, 11, 15, 22, 24, 27, 30, 31, 31, 41

Statistics Graphs and Charts (p104-108)

5) As well as counting the number of symbols on a pictogram, you need to check
one other thing before you can find a frequency. What's the other thing?

6) The numbers of students in different years at a village school
are shown in this table. Draw a bar chart to show this data.

School Year	7	8	9	10	11
Number of students	40	30	40	45	25

7) 125 boys and 125 girls were asked if they prefer Maths or Science.
74 of the boys said they prefer Maths, while 138 students altogether
said they prefer Science. How many girls said they prefer Science?
8) Draw a pie chart to represent the data in question 6.
9) Sketch graphs to show:
    a) weak positive correlation, b) strong negative correlation, c) no correlation
10) This stem and leaf diagram shows the speeds of the fastest serves of some tennis players.
Find the: a) fastest speed recorded   b) median speed   c) range of speeds

```
10 | 2 5
11 | 1 4 6
12 | 0 2 2
13 | 6 8
```
Key: 10 | 2 = 102 mph

Frequency Tables and Frequency Polygons (p109-112)

11) Explain how you would find the mode, median and mean of the data in a frequency table.

12) For this grouped frequency table showing the lengths of some pet alligators:
    a) find the modal class,
    b) find the class containing the median,
    c) estimate the mean.

Length (y, in m)	Frequency
$1.4 \leq y < 1.5$	4
$1.5 \leq y < 1.6$	8
$1.6 \leq y < 1.7$	5
$1.7 \leq y < 1.8$	2

13) The frequency polygons on the right show the times (t, in seconds)
taken to run 60 m by groups of Year 7 and Year 10 students.
Make one comment to compare the two sets of data.

Probability (p113-116)

14) What does a probability of 0 mean? What about a probability of ½?
15) I pick a random number between 1 and 50. Find the probability that my number is a multiple of 6.
16) If the probability of a spinner landing on red is 0.3,
what is the probability that it doesn't land on red?

HT means Heads on the first flip and Tails on the second.

17) I flip a fair coin twice.
    a) Complete this sample space diagram showing all the possible results.
    b) Use your diagram to find the probability of getting 2 Heads.

		Second flip	
		Heads	Tails
First flip	Heads		HT
	Tails		

18) Write down the formula for estimating how many times you'd expect something to happen in n trials.
19) I flip a fair coin 100 times. How many times would you expect it to land on Tails?
20) When might you need to use relative frequency to find a probability?

Answers

Section One

Page 2 — Calculating Tips
Q1 a) 11 **b)** 37 **c)** 3

Page 3 — Calculating Tips
Q1 0.149598822

Page 4 — Ordering Numbers and Place Value
Q1 a) One million, two hundred and thirty-four thousand, five hundred and thirty-one
b) Twenty-three thousand, four hundred and fifty-six
c) Three thousand, four hundred and two
d) Two hundred and three thousand, four hundred and twelve
Q2 56 421
Q3 9, 23, 87, 345, 493, 1029, 3004
Q4 0.008, 0.09, 0.1, 0.2, 0.307, 0.37

Page 5 — Addition and Subtraction
Q1 a) 171 cm **b)** 19 cm
Q2 1.72 litres

Page 6 — Multiplying by 10, 100, etc.
Q1 a) 1230 **b)** 3450 **c)** 9650
Q2 a) 48 **b)** 450 **c)** 180 000

Page 7 — Dividing by 10, 100, etc.
Q1 a) 0.245 **b)** 6.542 **c)** 0.00308
Q2 a) 1.6 **b)** 12 **c)** 5

Page 8 — Multiplying Without a Calculator
Q1 a) 336 **b)** 616 **c)** 832

Page 9 — Dividing Without a Calculator
Q1 a) 12 **b)** 12 **c)** 21
Q2 10 cm

Page 10 — Multiplying and Dividing with Decimals
Q1 a) 179.2 **b)** 6.12 **c)** 56.1
Q2 a) 56 **b)** 30 **c)** 705

Page 11 — Negative Numbers
Q1 a) $-6\,°C$ **b)** $-5\,°C$

Page 12 — Special Types of Number
Q1 a) 23 **b)** 125
Q2 If a is odd, then a^2 is odd (as odd × odd = odd). Then $a^2 - 2$ is always odd (as odd − even = odd).

Page 13 — Prime Numbers
Q1 61, 53, 47

Page 14 — Multiples, Factors and Prime Factors
Q1 a) 15 **b)** 13
Q2 $160 = 2 × 2 × 2 × 2 × 2 × 5$

Page 15 — LCM and HCF
Q1 36 **Q2** 12

Page 16 — Fractions, Decimals and Percentages
Q1 a) $\frac{2}{5}$ **b)** $\frac{1}{50}$ **c)** $\frac{77}{100}$
d) $\frac{111}{200}$ **e)** $\frac{28}{5}$
Q2 a) 57% **b)** $\frac{6}{25}$ **c)** 90%

Page 17 — Fractions
Q1 $\frac{4}{9}$ **Q2** 90 kg

Page 18 — Fractions
Q1 a) $\frac{12}{35}$ **b)** $\frac{3}{2}$
Q2 a) $\frac{37}{9}$ **b)** $5\frac{2}{3}$
Q3 a) $\frac{17}{32}$ **b)** $\frac{2}{3}$

Page 19 — Fractions
Q1 $\frac{11}{20}, \frac{5}{8}, \frac{7}{10}, \frac{3}{4}$
Q2 a) $\frac{3}{5}$ **b)** $\frac{1}{6}$

Page 20 — Fractions and Recurring Decimals
Q1
$$6\,\overline{\smash{\big)}\,1.1^10^40^40^40^40}$$
$$0.1\,6\,6\,6\,6$$
$1 ÷ 6 = 0.1666...$ so $\frac{1}{6} = 0.1\dot{6}$

Page 21 — Proportion Problems
Q1 56 p
Q2 a) 900 g **b)** 180 g

Page 22 — Proportion Problems
Q1 550 g, 342 g, 910 g

Page 23 — Percentages
Q1 549 ml **Q2** 27 %

Page 24 — Percentages
Q1 10% of 180 kg = 180 ÷ 10 = 18 kg
5% of 180 kg = 18 ÷ 2 = 9 kg
So 45% of 180 kg
= (4 × 10%) + 5%
= (4 × 18) + 9 = 72 + 9 = 81 kg
Q2 £209

Page 25 — Ratios
Q1 2 : 1
Q2 £3500, £2100, £2800

Page 26 — Rounding Off
Q1 a) 21.4 **b)** 0.06 **c)** 5.0
Q2 3.11

Page 27 — Rounding Off
Q1 a) 700 **b)** 15
c) 169 **d)** 82 000
Q2 4.8

Page 28 — Estimating Calculations
Q1 a) 20 **b)** 400
Q2 £40

Page 29 — Square Roots and Cube Roots
Q1 a) 14 **b)** 8 **c)** 7.5
Q2 a) 5 **b)** 10 **c)** 21

Page 30 — Powers
Q1 a) 43 **b)** 238.328 **c)** 10^6
Q2 a) 4^5 **b)** 7^3 **c)** q^8
Q3 $6^2 = 36$

Revision Questions — Section One
Q1 Twenty-one million, three hundred and six thousand, five hundred and fifteen.
Q2 2.09, 2.2, 3.51, 3.8, 3.91, 4.7
Q3 a) 882 **b)** 446 **c)** £4.17
Q4 a) £120 **b)** £0.50 = 50p
Q5 a) 1377 **b)** 26
c) 62.7 **d)** 0.35
Q6 a) −16 **b)** 7 **c)** 20
Q7 A square number is a whole number multiplied by itself.

Answers

The first ten are: 1, 4, 9, 16, 25, 36, 49, 64, 81 and 100.

Q8 41, 43, 47, 53, 59

Q9 The multiples of a number are its times table.
  a) 10, 20, 30, 40, 50, 60
  b) 4, 8, 12, 16, 20, 24

Q10 a) $210 = 2 \times 3 \times 5 \times 7$
  b) $1050 = 2 \times 3 \times 5 \times 5 \times 7$

Q11 a) 14 **b)** 40

Q12 a) i) $\frac{4}{100} = \frac{1}{25}$ **ii)** 4%

  b) i) $\frac{65}{100} = \frac{13}{20}$ **ii)** 0.65

Q13 To simplify a fraction you divide the top and bottom by the same number. To simplify as far as possible you keep dividing until they won't go any further.

Q14 a) 320 **b)** £60

Q15 a) $\frac{23}{8} = 2\frac{7}{8}$ **b)** $\frac{11}{21}$

  c) $\frac{25}{16} = 1\frac{9}{16}$ **d)** $\frac{44}{15} = 2\frac{14}{15}$

Q16 Recurring decimals have a pattern of numbers which repeats forever. You show a decimal is recurring by putting a dot above the digits at the start and end of the repeating part of the number (if only one digit is repeated then just put one dot above that digit).

Q17 £1.41

Q18 The 250 g tin is the best buy.

Q19 To find x as a percentage of y, divide x by y and then multiply by 100.

Q20 £60

Q21 The top costs £38.25 so Carl can't afford it.

Q22 5 : 8

Q23 600, 960, 1440

Q24 27 litres

Q25 a) 17.7 **b)** 6700
  c) 4 000 000

Q26 a) 100 **b)** 1400

Q27 a) 11 **b)** 4 **c)** 56 **d)** 10^4

Q28 a) 421.875 **b)** 4.8 **c)** 8

Q29 a) 1) When multiplying, add the powers.
     2) When dividing, subtract the powers.
     3) When raising one power to another, multiply the powers.
  b) 7^5

Section Two

Page 32 — Algebra — Simplifying Terms

Q1 a) $6a$ **b)** $7b$

Q2 $3x + 8y$

Page 33 — Algebra — Simplifying Terms

Q1 a) e^5 **b)** $18fg$

Q2 a) h^9 **b)** s^3

Page 34 — Algebra — Multiplying Out Brackets

Q1 a) $-18x + 12$ **b)** $3x^2 - 15x$

Q2 $2y + 19$

Page 35 — Algebra — Taking Out Common Factors

Q1 a) $7(3x - 2y)$ **b)** $x(x + 2)$

Q2 a) $6a(3 + 2a)$ **b)** $2r(2r - 11s)$

Page 36 — Solving Equations

Q1 a) $x = 6$ **b)** $x = 14$
  c) $x = 3$ **d)** $x = 15$

Page 37 — Solving Equations

Q1 $x = 5$

Q2 $y = 7$

Page 38 — Using Formulas

Q1 $Z = 39$

Page 39 — Making Formulas From Words

Q1 $C = 12d + 18$

Page 40 — Rearranging Formulas

Q1 $v = 3(u + 2)$ or $v = 3u + 6$

Q2 $d = \frac{c}{6} + 2$

Page 41 — Number Patterns and Sequences

Q1 Next term = 7. The rule is subtract 5 from the previous term.

Page 42 — Number Patterns and Sequences

Q1 a) $7n - 5$ **b)** 51
  c) No, as the solution to $7n - 5 = 63$ doesn't give an integer value of n.

Page 43 — Trial and Improvement

Q1 3.6 **Q2** 5.5

Page 44 — Inequalities

Q1 $n = -1, 0, 1, 2, 3, 4$

Q2 a) $x < 6$ **b)** $x \geq 3$

Revision Questions — Section Two

Q1 a) $3e$ **b)** $8f$

Q2 a) $7x - y$ **b)** $3a + 9$

Q3 a) m^3 **b)** $7pq$ **c)** $18xy$

Q4 a) g^{11} **b)** c^3

Q5 a) $6x + 18$ **b)** $-9x + 12$
  c) $5x - x^2$

Q6 $6x$

Q7 Putting in brackets (the opposite of multiplying out brackets).

Q8 a) $8(x + 3)$ **b)** $9(2x + 3y)$
  c) $5x(x + 3)$

Q9 a) $x = 7$ **b)** $x = 16$ **c)** $x = 3$

Q10 a) $x = 4$ **b)** $x = 2$ **c)** $x = 3$

Q11 $Q = 8$

Q12 $P = 7d + 5c$

Q13 2 hours

Q14 $v = \frac{W - 5}{4}$

Q15 a) 31, rule is add 7
  b) 256, rule is multiply by 4
  c) 19, rule is add two previous terms.

Q16 6n − 2

Q17 Yes, it's the 5th term.

Q18 A way of finding an approximate solution to an equation by trying different values in the equation.

Q19 $x = 4.1$

Q20 $x = 3.7$

Q21 a) x is greater than minus seven.
  b) x is less than or equal to 6.

Q22 $k = 1, 2, 3, 4, 5, 6, 7$

Q23 a) $x < 10$ **b)** $x > 14$ **c)** $x \geq 3$

Q24 $x \leq 7$

Section Three

Page 46 — Coordinates and Midpoints

Q1 a)
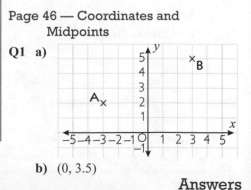

  b) (0, 3.5)

Answers

Page 47 — Straight-Line Graphs
Q1 **a), b) and c)**

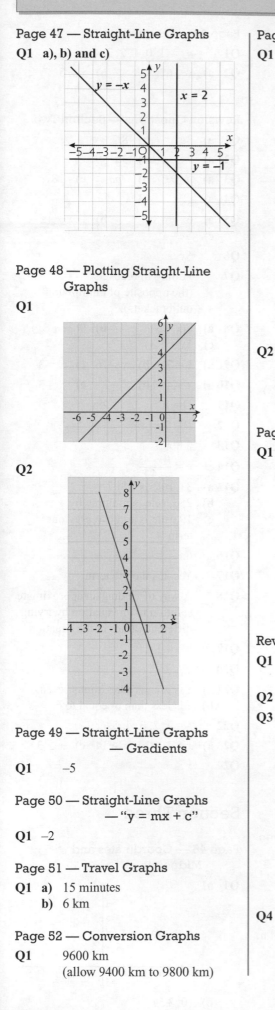

Page 48 — Plotting Straight-Line Graphs
Q1

Q2

Page 49 — Straight-Line Graphs — Gradients
Q1 –5

Page 50 — Straight-Line Graphs — "y = mx + c"
Q1 –2

Page 51 — Travel Graphs
Q1 **a)** 15 minutes
b) 6 km

Page 52 — Conversion Graphs
Q1 9600 km
(allow 9400 km to 9800 km)

Page 53 — Real-Life Graphs
Q1 **a)**

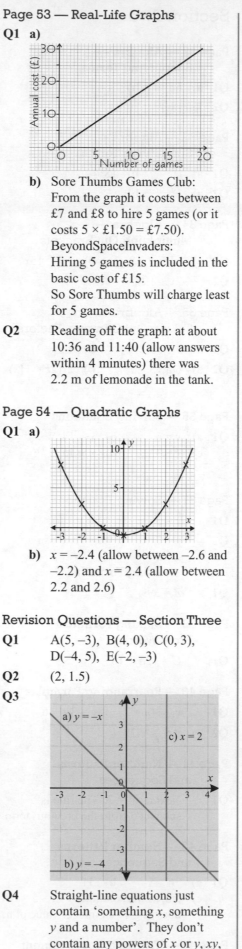

b) Sore Thumbs Games Club:
From the graph it costs between £7 and £8 to hire 5 games (or it costs 5 × £1.50 = £7.50).
BeyondSpaceInvaders:
Hiring 5 games is included in the basic cost of £15.
So Sore Thumbs will charge least for 5 games.

Q2 Reading off the graph: at about 10:36 and 11:40 (allow answers within 4 minutes) there was 2.2 m of lemonade in the tank.

Page 54 — Quadratic Graphs
Q1 **a)**

b) $x = -2.4$ (allow between –2.6 and –2.2) and $x = 2.4$ (allow between 2.2 and 2.6)

Revision Questions — Section Three
Q1 A(5, –3), B(4, 0), C(0, 3), D(–4, 5), E(–2, –3)

Q2 (2, 1.5)

Q3

Q4 Straight-line equations just contain 'something x, something y and a number'. They don't contain any powers of x or y, xy, $1/x$ or $1/y$.

Q5 E.g.

x	0	1	2
y	3	5	7

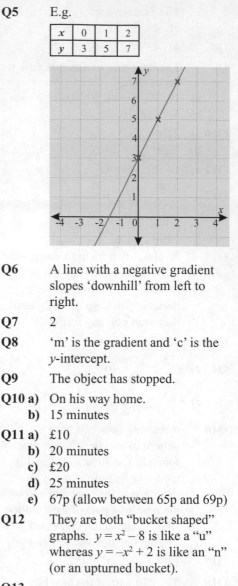

Q6 A line with a negative gradient slopes 'downhill' from left to right.

Q7 2

Q8 'm' is the gradient and 'c' is the y-intercept.

Q9 The object has stopped.

Q10 a) On his way home.
b) 15 minutes

Q11 a) £10
b) 20 minutes
c) £20
d) 25 minutes
e) 67p (allow between 65p and 69p)

Q12 They are both "bucket shaped" graphs. $y = x^2 - 8$ is like a "u" whereas $y = -x^2 + 2$ is like an "n" (or an upturned bucket).

Q13

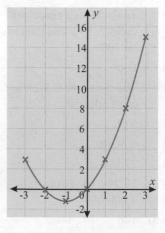

$x = -2.7$ (allow between –2.6 and –2.9) and $x = 0.7$ (allow between 0.6 and 0.9).

Answers

Section Four

Page 56 — Symmetry
Q1 a) E.g.

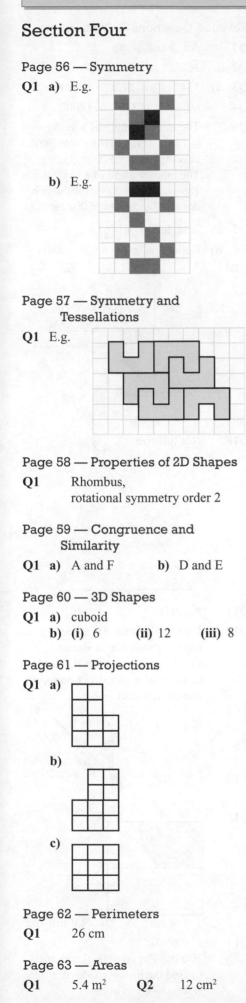

b) E.g.

Page 57 — Symmetry and Tessellations
Q1 E.g.

Page 58 — Properties of 2D Shapes
Q1 Rhombus, rotational symmetry order 2

Page 59 — Congruence and Similarity
Q1 a) A and F **b)** D and E

Page 60 — 3D Shapes
Q1 a) cuboid
b) (i) 6 **(ii)** 12 **(iii)** 8

Page 61 — Projections
Q1 a)

b)

c)

Page 62 — Perimeters
Q1 26 cm

Page 63 — Areas
Q1 5.4 m^2 **Q2** 12 cm^2

Page 64 — Areas
Q1 3 tins

Page 65 — Circles
Q1 A = radius, B = chord, C = tangent

Page 66 — Circle Questions
Q1 10 603 cm^2 (nearest cm^2)

Page 67 — Volume
Q1 360 cm^3

Page 68 — Nets and Surface Area
Q1 E.g.

Page 69 — Nets and Surface Area
Q1 184 cm^2

Revision Questions — Section Four
Q1 H: 2 lines of symmetry, rotational symmetry order 2
Z: 0 lines of symmetry, rotational symmetry order 2
T: 1 line of symmetry, rotational symmetry order 1
N: 0 lines of symmetry, rotational symmetry order 2
E: 1 line of symmetry, rotational symmetry order 1
X: 4 lines of symmetry, rotational symmetry order 4
S: 0 lines of symmetry, rotational symmetry order 2

Q2 E.g.

Q3 No

Q4 2 angles the same, 2 sides the same, 1 line of symmetry, no rotational symmetry.

Q5 2 lines of symmetry, rotational symmetry order 2

Q6 2 pairs of equal (parallel) sides, 2 pairs of equal angles, no line of symmetry, rotational symmetry order 2.

Q7 Congruent shapes are exactly the same size and same shape. Similar shapes are the same shape but different sizes.

Q8 a) D and G
b) C and F

Q9 a) faces = 5, edges = 8, vertices = 5
b) faces = 2, edges = 1, vertices = 1
c) faces = 5, edges = 9, vertices = 6

Q10 The view from directly above an object.

Q11 Front: Side:

Plan:

Q12 21 cm
Q13 32 cm^2
Q14 Area = ½(a + b) × h
Q15 36 cm^2
Q16 52 cm^2
Q17 9 mm
Q18 Area = 153.94 cm^2 (2 d.p.)
Circumference = 43.98 cm (2 d.p.)
Q19 E.g.

Arc
Sector
Segment

Q20 Area = 7.07 cm^2 (2 d.p.)
Perimeter = 10.71 cm (2 d.p.)
Q21 Volume = $\pi r^2 h$
Q22 360 cm^3
Q23 150 cm^2
Q24 125.7 cm^2 (1 d.p.)

Section Five

Page 71 — Lines and Angles
Q1 a) Acute
b) e.g. 190° (any angle more than 180° but less than 360°)

Answers

Page 72 — Measuring & Drawing Angles

Q1

227°

Q2 119° (allow between 118°-120°)

Page 74 — Five Angle Rules

Q1 108°

Page 76 — Parallel Lines

Q1 123°

Page 77 — Polygons and Angles

Q1 144° **Q2** Pentagon

Page 78 — Polygons, Angles and Tessellations

Q1 a) 900°

b) Interior angle of regular heptagon = 900 ÷ 7 = 128.571...°
This does not divide into 360° exactly, so regular heptagons don't tessellate.

Page 79 — Transformations

Q1 Translation by the vector $\begin{pmatrix} -8 \\ -1 \end{pmatrix}$.

Q2

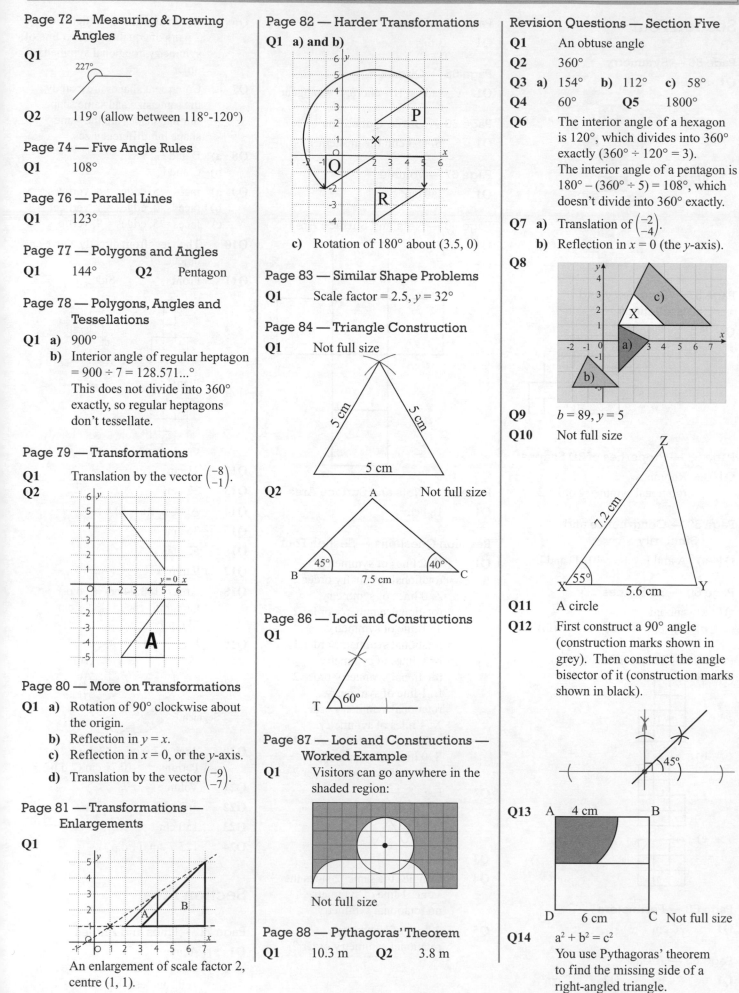

Page 80 — More on Transformations

Q1 a) Rotation of 90° clockwise about the origin.

b) Reflection in $y = x$.

c) Reflection in $x = 0$, or the y-axis.

d) Translation by the vector $\begin{pmatrix} -9 \\ -7 \end{pmatrix}$.

Page 81 — Transformations — Enlargements

Q1

An enlargement of scale factor 2, centre (1, 1).

Page 82 — Harder Transformations

Q1 a) and b)

c) Rotation of 180° about (3.5, 0)

Page 83 — Similar Shape Problems

Q1 Scale factor = 2.5, $y = 32°$

Page 84 — Triangle Construction

Q1 Not full size

5 cm, 5 cm, 5 cm

Q2 Not full size

A, 45°, 40°, B, 7.5 cm, C

Page 86 — Loci and Constructions

Q1

T, 60°

Page 87 — Loci and Constructions — Worked Example

Q1 Visitors can go anywhere in the shaded region:

Not full size

Page 88 — Pythagoras' Theorem

Q1 10.3 m **Q2** 3.8 m

Revision Questions — Section Five

Q1 An obtuse angle

Q2 360°

Q3 a) 154° **b)** 112° **c)** 58°

Q4 60° **Q5** 1800°

Q6 The interior angle of a hexagon is 120°, which divides into 360° exactly (360° ÷ 120° = 3).
The interior angle of a pentagon is 180° − (360° ÷ 5) = 108°, which doesn't divide into 360° exactly.

Q7 a) Translation of $\begin{pmatrix} -2 \\ -4 \end{pmatrix}$.

b) Reflection in $x = 0$ (the y-axis).

Q8

Q9 $b = 89$, $y = 5$

Q10 Not full size

Z, 7.2 cm, 55°, X, 5.6 cm, Y

Q11 A circle

Q12 First construct a 90° angle (construction marks shown in grey). Then construct the angle bisector of it (construction marks shown in black).

45°

Q13 A, 4 cm, B, D, 6 cm, C Not full size

Q14 $a^2 + b^2 = c^2$
You use Pythagoras' theorem to find the missing side of a right-angled triangle.

Answers

Q15 4.7 m   **Q16** 14.5 cm
Q17 5

Section Six

Page 90 — Metric and Imperial Units

Q1 E.g.

	Metric	Imperial
Speed of a train	km/h	mph
Weight of an apple	grams	ounces
Volume of a bottle of milk	litres	pints

Page 91 — Converting Units

Q1 0.18 kg   **Q2** 1237 ml

Page 92 — More Conversions

Q1 225.3 cm²   **Q2** 6.25 mph

Page 93 — Reading Scales

Q1 75 mph   **Q2** 3.2 °C

Page 94 — Rounding and Estimating Measurements

Q1 a) 4.2 cm
b)

M ——————×—————— N

Q2 Answer in range 0.8 m – 1 m.

Page 95 — Reading Timetables

Q1 99 minutes
Q2
19:12 Leave house
19:22 Catch bus at Market Street
19:31 Arrive at shops

Page 96 — Compass Directions and Bearings

Q1 310°   **Q2** 035°

Page 97 — Maps and Scale Drawings

Q1 9 miles   **Q2** 4.5 cm

Page 98 — Maps and Scale Drawings

Q1 1 cm represents 200 m

E.g.
Woodland
400 m
600 m

Q2

Cinema ×
N
2.5 cm
35°
House
1 cm = 2 miles

Page 99 — Speed

Q1 15 minutes
Q2 Rachel's average speed is 4.5 ÷ 1.5 = 3 km/h, Tyler's is 7 ÷ 2 = 3.5 km/h, so Tyler has the fastest average speed.

Revision Questions — Section Six

Q1 1 m = 100 cm
1 litre = 1000 cm³
1000 m = 1 km
1 litre ≈ 1.75 pints
1 mile ≈ 1.6 km
1 foot ≈ 30 cm

Q2 1) Find the conversion factor
2) Multiply and divide by it
3) Choose the common-sense answer.

Q3 65 kg

Q4 Use the conversion factor twice for area and three times for volume.

Q5 0.12 m³

Q6 To work out what a small gap represents, divide the size of the large gaps between numbers by the number of small gaps between numbers.

Q7 5.2 cm

Q8 5.6 cm

A ——————————×—— B

Q9 Answer in range 1 m – 1.4 m

Q10 a) 16:20   **b)** 7.52 am

Q11 151 minutes

Q12 a) 12:18   **b)** 11:58

Q13 From, North line, Clockwise

Q14 Multiply by the map scale to go from map distance to real life. Divide by the map scale to go from real life to map distance. The scale needs to be in the form 1 cm = ...

Q15 Plan should be a rectangle that's 20 cm long and 12 cm wide.

Q16 110°

Q17

N
Port P
1.5 cm   230°
×
Port Q

Q18

D / S × T

Q19 24 m

Section Seven

Page 101 — Collecting Data

Q1 Two from, e.g: sample too small, one city centre not representative of the whole of Britain, only done in one particular place.

Page 102 — Collecting Data

Q1 **1st question:**
Question should include a time frame, e.g. "How much do you spend on food each week?" Include at least 3 non-overlapping response boxes, covering all possible answers, e.g. 'less than £20', '£20 to £40', 'over £40' etc.
2nd question:
Include at least 3 non-overlapping response boxes, covering all possible answers, e.g. '0-1', '2-3', etc.
3rd question:
Make the question fair, e.g. "Which do you prefer, potatoes or cabbage?"
Response boxes should cover all answers, e.g. 'Potatoes', 'Cabbage', or 'Don't know'.
4th question:
Limit the number of possible answers, e.g. "Choose your favourite food from the following options." Include at least 3 non-overlapping response boxes.

Page 103 — Mean, Median, Mode and Range

Q1 a) Mean = 11.2, median = 12, mode = 12, range = 9
b) The first set of scores has a higher mean, so those scores are generally higher. The second set has a bigger range, so those scores are more spread out.

Page 104 — Pictograms and Bar Charts

Q1 a) 3
b)

Rock	◐
Blues	●●
Opera	●●◐
Jazz	●◐

Page 105 — Two-Way Tables

Q1

	Walk	Car	Bus	Total
Male	15	21	13	49
Female	18	11	22	51
Total	33	32	35	100

Answers

Page 106 — Pie Charts

Q1

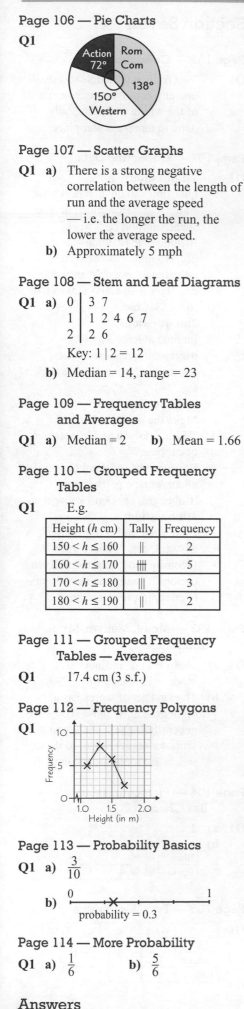

Action 72°
Rom Com 138°
150° Western

Page 107 — Scatter Graphs

Q1 a) There is a strong negative correlation between the length of run and the average speed — i.e. the longer the run, the lower the average speed.

b) Approximately 5 mph

Page 108 — Stem and Leaf Diagrams

Q1 a)
```
0 | 3 7
1 | 1 2 4 6 7
2 | 2 6
```
Key: 1 | 2 = 12

b) Median = 14, range = 23

Page 109 — Frequency Tables and Averages

Q1 a) Median = 2   **b)** Mean = 1.66

Page 110 — Grouped Frequency Tables

Q1 E.g.

Height (h cm)	Tally	Frequency
$150 < h \le 160$	||	2
$160 < h \le 170$	||||	5
$170 < h \le 180$	|||	3
$180 < h \le 190$	||	2

Page 111 — Grouped Frequency Tables — Averages

Q1 17.4 cm (3 s.f.)

Page 112 — Frequency Polygons

Q1

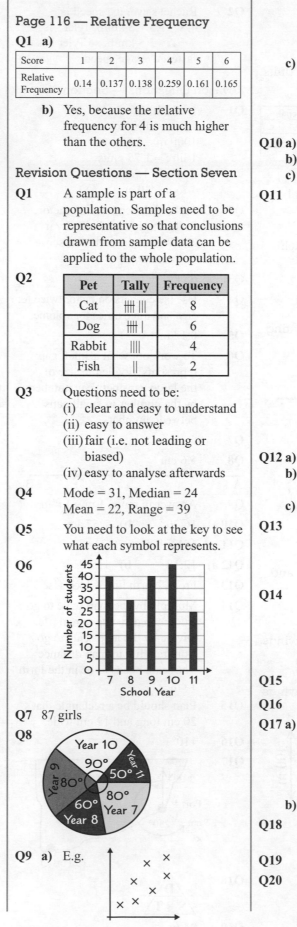

Page 113 — Probability Basics

Q1 a) $\frac{3}{10}$

b)
```
0 ——————X———————— 1
```
probability = 0.3

Page 114 — More Probability

Q1 a) $\frac{1}{6}$ **b)** $\frac{5}{6}$

Page 115 — Expected Frequency

Q1 60 times

Page 116 — Relative Frequency

Q1 a)

Score	1	2	3	4	5	6
Relative Frequency	0.14	0.137	0.138	0.259	0.161	0.165

b) Yes, because the relative frequency for 4 is much higher than the others.

Revision Questions — Section Seven

Q1 A sample is part of a population. Samples need to be representative so that conclusions drawn from sample data can be applied to the whole population.

Q2

Pet	Tally	Frequency
Cat	|||| |||	8
Dog	|||| |	6
Rabbit	||||	4
Fish	||	2

Q3 Questions need to be:
(i) clear and easy to understand
(ii) easy to answer
(iii) fair (i.e. not leading or biased)
(iv) easy to analyse afterwards

Q4 Mode = 31, Median = 24
Mean = 22, Range = 39

Q5 You need to look at the key to see what each symbol represents.

Q6

Number of students:
Year 7: 40
Year 8: 30
Year 9: 40
Year 10: 45
Year 11: 25

Q7 87 girls

Q8

Year 10: 90°
Year 11: 50°
Year 7: 80°
Year 8: 60°
Year 9: 80°

Q9 a) E.g.

b) E.g.

c) E.g.

Q10 a) Fastest speed = 138 mph

b) Median = 118 mph

c) Range = 36 mph

Q11 To find the mode: find the category with the highest frequency.
To find the median: work out the position of the middle value, then count through the frequency column to find the category it's in.
To find the mean: add a third column to the table showing the values in the first column (x) multiplied by the values in the frequency column (f). Work out the mean by dividing the total of the 3rd column (total of $f \times x$) by the total of the frequency column (total of f).

Q12 a) Modal class is: $1.5 \le y < 1.6$.

b) Class containing median is: $1.5 \le y < 1.6$

c) Estimated mean = 1.58 m (2 d.p.)

Q13 There are lots of things you could say — e.g. far more Year 10 students than Year 7 students took less than 10 seconds to run 60 m.

Q14 A probability of 0 means something will never happen.
A probability of ½ means something is as likely to happen as not.

Q15 $\frac{4}{25}$

Q16 0.7

Q17 a)

		Second flip	
		Heads	Tails
First flip	Heads	HH	HT
	Tails	TH	TT

b) $\frac{1}{4}$

Q18 Expected frequency = probability × number of trials

Q19 50 times

Q20 When you can't tell the probability of each result 'just by looking at it'. / When a dice, coin or spinner is biased.

Answers

Index

Index